CHEMICAL BONDS AND BOND ENERGY

Second Edition

This is Volume 21 of
PHYSICAL CHEMISTRY
A Series of Monographs

Editor: ERNEST M. LOEBL, *Polytechnic Institute of New York*

A complete list of titles in this series appears at the end of this volume.

CHEMICAL BONDS AND BOND ENERGY

Second Edition

R. T. SANDERSON

Department of Chemistry
Arizona State University
Tempe, Arizona

ACADEMIC PRESS   New York   San Francisco   London   1976

A SUBSIDIARY OF HARCOURT BRACE JOVANOVICH, PUBLISHERS

ACADEMIC PRESS, INC.
111 Fifth Avenue, New York, New York 10003

United Kingdom Edition published by
ACADEMIC PRESS, INC. (LONDON) LTD.
24/28 Oval Road, London NW1

Library of Congress Cataloging in Publication Data

Sanderson, Robert Thomas, (date)
    Chemical bonds and bond energy.

    (Physical chemistry, a series of monographs ;   )
    Includes bibliographical references.
    1.   Chemical bonds.   I.   Title.   II.   Series.
QD461.S33   1976   541'.224   75-26353
ISBN 0–12–618060–1

Contents

Preface   ix

Preface to First Edition   xi

ONE / The Delightful Science of Chemistry: A Preview of a New
    View

    The Periodic Law   2
    Potassium and Chlorine   4
    Potassium Chloride   7
    The Burning of Carbon   16
    Contributing Bond Energy and Bond Dissociation Energy   22
    Summary and Comments   24

TWO / Significant Atomic Properties of the Chemical Elements

    Atomic Structure   26
    Atomic Properties   31

THREE / Selected Values of Atomic Properties

    Introduction   52
    Major Group Elements   52
    Other Major Group Elements   64
    Transitional Elements   65
    Summary   66

FOUR / The Physical States of Nonmetals

    The Monatomic Gaseous Elements of Group M8   69
    The Halogens   71
    The Chalcogens   71
    Elements of Group M5   73
    Elements of Group M4   73

FIVE / Polar Covalence I: Electronegativity Equalization, Partial
       Charge, and Bond Length

The Principle of Electronegativity Equalization   75
Partial Charge   77
The Radii of Partially Charged Atoms, and Their Sum as Bond
    Length   79

SIX / Polar Covalence II: The Calculation of Polar Bond Energy

The Need for Bond Energy Calculation   95
The Calculation of Heteronuclear Bond Energy   97
Nonmolecular Solids   105

SEVEN / Chemical Combinations of Hydrogen

The Special Nature of Hydrogen   111
Spectroscopic (Diatomic) Hydrides   114
Divalent Hydrogen   114
The Periodicity of Hydrogen Chemistry   116

EIGHT / The Chemical Behavior of Oxygen

Bonding by Oxygen   123
Oxygen with Partial Charge   124
Bond Energies in Volatile Oxides   125
Solid Oxides   134
Hydroxides, Oxyacids   138

NINE / Halide Chemistry

Partial Charge on Halogen and Halide Properties   142
Bonds and Bond Energies in Halides   146
Atomization of Oxyhalides   155

TEN / Bond Dissociation

The Interrelationship between Contributing Bond Energy and
    Bond Dissociation Energy   156
Reorganizational Energies of Radicals   158
What Is the Evidence?   160
Speculative Explanations of Reorganizational Energy   171
General Applications of BDE   172
Coordination in Molecular Addition Compounds   173

ELEVEN / Chemical Bonds in Organic Compounds

The Nature of the Problem   177
Contributing Bond Energies   182
Standard Bond Energies   184
More about Contributing Bond Energy   185
Some Applications of Reorganizational Energies   186
Organometallic Compounds   188
Cohesive Energy   189

TWELVE / Summary: Application to the Oxidation of Ethane

The Nature of Ethane   192
Oxidation Step One: Ethanol   194
Oxidation Step Two: The Second Hydroxyl   198
Oxidation Step Three: Acetic Acid   201
Oxidation Completion: Carbon Dioxide and Water   203
Conclusion   204

Table of Logarithms   205

Index   207

Preface

The first edition of this book contained a fully detailed account of bond energy calculation based on fundamental properties of atoms. On the basis of a simple concept of polar covalence, it provided a means of accounting quantitatively for heats of formation and reaction. Thus the book did much to reveal the beautiful cause-and-effect relationship between the qualities of atoms and the physical and chemical properties of their chemical combinations. It also presented new and valuable insights into the nature of bond multiplicity and the "lone pair bond weakening effects" that appears so significant in the chemistry of elements of the periodic groups M5, M6, and M7.

Regretfully but deliberately omitted from the first edition was any explanation of the well-known fact that the bond energy mentioned above is usually not at all the same as the energy required to break the bond if there is more than one bond per molecule. The two kinds of bond energy may be distinguished by calling the former the *contributing bond energy* (CBE) and the latter, as usual, the *bond dissociation energy* (BDE). Much more has now been learned which is included in this second edition. The CBE is closely related to the BDE, being an important part of it, the remainder being the reorganizational energy (E_R) of the radicals created when the bond is broken: $BDE = CBE + E_R$. E_R has been evaluated for each of many common radicals, as detailed herein. These values are useful if not indispensable in the interpretation of important phenomena such as reaction mechanisms and product distribution, of special interest to organic chemists but also to inorganic chemists.

Further research on bond lengths, successful for binary compounds, is reported herein, along with new material on bond energies in solids and molecular addition compounds.

The challenge of holding this edition to a reasonable size was met by eliminating some of the details of individual bond energy calculations and finding more efficient means of presenting the information. The book has been rewritten almost completely. The average reader should still be able to test thoroughly any of the material in this book or from elsewhere, using the basic information contained herein. The author will welcome inquiries or suggestions from readers.

   In a world, even in a scientific world, where censorship remains an approved means of expressing disagreement, it is refreshing to be able to acknowledge the existence of openmindedness. In particular, I should like to express my deep appreciation to Professor Leland C. Allen of Princeton and Professor Robert G. Parr of the University of North Carolina for their kind and generous encouragement. Finally, my thanks to my nearly nonagenarian mother, who has no need of understanding this book to know it is a good one.

Preface to First Edition

The ultimate goal of theoretical chemistry is the attainment of so thorough an understanding of atoms that their complete behavior under all conditions of chemical interest can be predicted, together with the physical and chemical properties of all substances and their mixtures. In other words, the cause-and-effect relationship between the nature of atoms and the nature of their combinations should become perfectly revealed.

One of the central problems in the pursuit of this unattainable yet irresistibly challenging goal has been to understand the nature of chemical bonds. To understand bonds, one must be able to calculate their energies. This book tells how. It reports the first generally successful calculation of more than 850 kinds of bonds in more than 500 compounds. Such calculations provide fascinating new insights regarding the nature of bonds. In turn, these insights permit the first successful explanations of many previously puzzling phenomena, which are also included. I find these ideas immensely helpful in the classroom, and hope my fellow teachers will share this experience. The work should be equally useful to students and practicing chemists.

The advent of quantum mechanics stirred high hopes that the whole of chemical science could be created from fundamental theory. An abysmal gap soon became evident, however, between principle and practice. Quantum mechanics has been of indispensable assistance in the development of modern chemical theories of atomic structure, and atomic and molecular spectroscopy, and in many other areas. But even the heaviest artillery of quantum mechanics, brought to fullest effect over a period of forty years through the medium of modern computers, has scarcely dented the problem of bond energy calculation.

There are two good reasons for this failure. One is the immense complexity of the practically insoluble problem of calculating all the interactions among all the component particles of an atom. The other is the fact that the energies of interactions among atoms are usually far smaller than the total energies of the atoms. The logical calculation of bonding energy as a difference between the total energy of a molecule and the total energy of its atoms is therefore subject to the difficulty of obtaining accurately very small differences between very large values. Compare, for example, the energies of two oxygen atoms (obtained as the sum of the successive ionization energies), roughly 94,000 kcal per mole, with the O_2 bond energy of 119. Even very good approximate solutions of the many-body problem could hardly be expected to provide reliable bond energies.

The approach I have taken, therefore, over the past twenty years, has been to accept the findings of quantum mechanics to the limit of their usefulness, and then deliberately to avoid the insoluble many-particle problem by attempting to identify those qualities of an atom which in a sense summarize, or are the resultant of, all its interelectronic and electron-nuclear interactions. I have now identified these qualities as the covalent radius, the electronegativity, and the homonuclear bond energy, and have shown how the latter two are interrelated and can be obtained one from the other. These atomic properties, plus the bond length, are the basic data for bond energy calculations as described herein.

This work has revealed many questions needing answer, which I hope many readers will become interested in pursuing.

For financial support, I am indebted to the University of Iowa and especially to Arizona State University for having provided me with a steady salary and comfortable working conditions. For their moral support I am grateful to my wife Bernice and my son Bob, and to my respected colleagues Dr. LeRoy Eyring, Dr. Sheng Lin, and Dr. Paul Stutsman, who have sympathetically strengthened my philosophical endurance of the frustrations of frequent opposition.

None of this work would have been possible without the help of contributions from both experimental and theoretical chemists far too numerous to acknowledge individually, but nonetheless deeply appreciated. Their data and ideas have been a constant source of inspiration.

ONE

The Delightful Science
of Chemistry:
A Preview of a New View

This chapter allows a sort of quick peek, or sneak preview, into matters that will be presented in considerable detail throughout the remainder of the book. Whether the view will be considered new must depend somewhat on what is meant by view, on what is meant by new, and on what may be new to the viewer. One may sit on a hilltop and watch the valley below. Initially it is completely covered with low-hanging clouds that obscure all but the grossest features of the valley. Slowly and gradually the clouds thin, revealing a small detail here, another there. Hints of scenic continuity begin to be recognized, but not until a substantial portion of the cloud cover has been dissipated can one begin to perceive what is really there. Step by step, one's imagination of the unseen becomes reinforced by reality, or proven absurd, by revelation of additional areas previously covered. And finally the last wisp of vapor becomes dispersed, and the whole valley lies appealingly before one's eyes. At what point in the disappearance of the clouds does the valley appear?

The fundamental nature of chemistry has lain under such a blanket of fog for years and years. The fog persists to this day concealing what one assumes to be the full beauty and exquisite system of chemical change and of physical and chemical properties. Over the years, spots of cloud have worn thin or been removed here and there allowing clear perception of portions of the landscape, but the full coherence of the scene still remains interrupted by areas of undissipated aerosol. Shall we withhold our enjoyment until the last vestige of fog has disappeared? Or shall we take pleasure in it now, letting our imaginations tentatively fill in those areas still blotted from our view? I think we should reserve our highest admiration for the impossible but, meanwhile, appreciate to the fullest that which we have.

For me, the fog has been slowly clearing throughout 45 years as a student of chemistry. Although it still clings with tantalizing tenacity to many mysterious places, sufficient of the whole has been revealed to allow, for the simple mind at least, a more satisfying understanding of chemistry than has ever before been possible. I have ample reason to suspect that their preoccupation with other important and fascinating matters has thus far kept my fellow chemists from developing a full appreciation of the rather remarkable progress toward understanding that is described in this book. I hope that the contents of this chapter may stimulate your curiosity, even whet your appetite, to examine the remainder of the book with a critical but open-minded attitude.

Two examples are our practical limit here. With only two we cannot begin to be comprehensive, but it is possible to become aware of some of the more interesting and exciting implications of this work. Let us therefore engage in a fairly detailed analysis of some of the more relevant chemistry of two common compounds, potassium chloride and carbon dioxide. In preparation for that, we may need reminding of the periodic law.

THE PERIODIC LAW

The basis of the periodic law is generally considered to be the **periodicity of electronic configurations** that characterizes the build-up of the chemical elements in succession of increasing atomic number. Only one step less fundamental, and at least equally important, is the effect of electronic configuration on the **extent to which nuclear charge can be felt.** The effectiveness of the positive charge of the nucleus, as sensed at the periphery of the atom, is called the **effective nuclear charge.** All that the *structure* tells us about bonding is which electrons are so located that they have the possibility of being shared with another atom and which orbitals have vacancies that might accommodate electrons of other atoms. *Effective nuclear charge,* however, determines *how readily* the electrons of the atom might be shared by another atom and *how strongly* the electrons of another atom might be attracted. Taken together, the atomic structure and the effective nuclear charge are the principal factors that govern the behavior of atoms when they come into contact. They consequently tend strongly to predetermine the results of interatomic interactions. These results include, of course, the nature of the physical and chemical properties of any compounds that are formed.

Consider the changes that occur from left to right across the major group elements within a period of the periodic table. The number of outermost electrons changes from 1 to 8. The number of orbital vacancies correspondingly decreases from 7 to 0. To the significant extent that the number of covalent bonds an atom can form is determined by the number of possible half-filled outermost orbitals, then for the first four groups the number of bonds, or valence, is limited by the number of outermost *electrons,* but the valence of the last four groups is limited by the number of outermost *vacancies*: 1–2–3–4–3–2–1–0. Broadly speaking, the question of whether like atoms will join together through covalence or by metallic bonding is largely a function of the ratio of outer electrons to outer vacancies. When the ratio is less than one, all the elements are metals with the single exception of the maverick boron. The small size of the boron atom seems to involve holding the outermost electrons too tightly for effective delocalization of the type characteristic of metals. When the ratio is greater than one, some metallic properties may be observed, for reasons requiring no discussion here, but most such elements are clearly nonmetals.

The outermost principal quantum level of an atom differentiates electrons from those lying in lower levels in the following manner. Whereas *underlying* electrons are quite effective in shielding the nuclear charge on an almost one-to-one basis, the *outermost* electrons appear too busy keeping out of each other's way to intervene between one another and the nucleus. Consequently they are very ineffective in blocking off nuclear charge, being roughly only about one-third efficient. This means that adding one positive charge to the nucleus while adding one electron to the outermost level produces an increase of about two-thirds of a protonic charge in the effective nuclear charge. Therefore filling of the outermost shell from one to eight electrons, in the building up of the chemical elements, is accompanied by a steady **increase** in effective nuclear charge.

The size of an electronic cloud surrounding an atomic nucleus, vaguely though this size must be defined, is a function of a balance between the interelectronic repulsions and the attractions between electrons and nucleus. Increasing effective nuclear charge must therefore cause contraction of the cloud, and the atomic radius **decreases** as the number of outermost electrons is increased from one to eight.

The electronegativity of an atom is proportional to the coulombic force between the effective nuclear charge and an electron in an outermost orbital. Therefore increasing the number of outermost electrons must **increase** the electronegativity, since it increases the effective nuclear charge and diminishes the distance over which it must be effective.

Ionization energies of the outermost electrons are of course influenced by orbital type and whether the electron is paired or unpaired in its orbital. But the general trend **upward** from left to right is the expected result of a steadily increasing effective nuclear charge.

Electron affinities are likewise determined by the effective nuclear charge and the distance over which it must act. A trend of **increasing** electron affinity is therefore expected as the number of electrons in the outermost level increases.

In summary, the numerical valence of an atom is determined by the electronic configuration; but the characterization as metal or nonmetal, oxidizing or reducing agent, and the general nature of its combinations with other atoms, both like and different, all reflect the extent to which the nuclear charge is able to control the conditions of both the outermost electrons and the outermost vacancies. With these principles in mind, it is then possible to examine the atoms of specific elements with a reasonably reliable preconceived concept of *what* these atoms must be like and *why* this should be so.

POTASSIUM AND CHLORINE

The central theme of this analysis is that **the properties of compounds must be predetermined by the nature of the atoms which compose them.** In that sense we should be able to predict, or at least enjoy the comfortable feeling, that the properties of atoms are responsible for the properties of compounds, and a sufficient understanding of atoms should lead logically to an adequate understanding of compound properties.

Potassium

One glance at the electronic configuration of potassium, as abbreviated by 2–8–8–1, should suffice to inform us that there is only one electron, and therefore seven vacancies, in the outermost principal quantum level of the potassium atom. The preceding element, argon 2–8–8, exhibits no appreciable effective nuclear charge from the viewpoint of using orbitals in the fourth principal level, for argon is essentially inert. In order to persuade this atom to accept an electron and hold on to it, it is necessary to increase its nuclear charge by one. Even so, this charge is largely blocked by the very symmetrical electronic cloud that surrounds it, so that the added electron (which makes the atom into one of potassium) is not very strongly held. This should imply several related bits of information about potassium atoms:

(1) The atoms should be relatively large, in fact the largest of all atoms in the period that potassium begins. We have experimental evidence of this size in the bond length observed for the K_2 molecule. This length, 3.92 Å, implies an effective radius in the bond direction of 1.96 Å for the potassium atom. Other evidences of relatively large size are to be found in consideration of the solid element, as will be discussed shortly.

(2) The atoms do not hold the outermost electron very strongly. In fact, the ionization energy of potassium is only 102 kpm (kilocalories per mole), which is lower than for any other elements except rubidium, cesium, and francium.

(3) If an atom cannot hold its own electrons tightly, then it certainly cannot be expected to attract outside electrons strongly. In an atom of potassium the effective nuclear charge is small and it is required to operate over a relatively large distance, which means that the electronegativity of the atom is very low. A value of 0.42 has been determined from a consideration of the average density of the electronic sphere, which shows on the average fewer electrons per unit volume of the sphere than in any other elements except the heavier ones of this group.

(4) If an argon atom cannot acquire an extra electron favorably, then certainly an atom of potassium cannot be expected to do so. This is equivalent to saying that the electron affinity must be energy absorbed, not evolved.

We are now ready to consider bringing potassium atoms together. Each has one outermost half-filled orbital and thus the requisites for the formation of one single covalent bond. It is therefore easy to predict with assurance that the covalency of potassium cannot exceed one and that two potassium atoms might join together by a single covalent bond to form a diatomic molecule K_2, fully using the covalent capacity of each atom. Since bonding forces involve attractions that increase with decreasing distance and since the potassium atoms are relatively large, the large distance would be expected to correspond to weak bonding. Furthermore, since the bonding results from the mutual attraction of both nuclei for the same two shared electrons and the low electronegativity shows that neither nucleus can attract such electrons very strongly, we can predict weak bonding for this reason also. The dissociation energy of the K_2 molecule has been measured to have the very low value of 13.2 kpm. In fact, this value can be calculated from a simple linear relationship that occurs between homonuclear bond energy and nonpolar radius and the electronegativity: $E = CrS$. Electronegativity is proportional to the coulomb *force* between the effective nuclear charge and an electron at the distance of the covalent radius. Homonuclear bond energy is proportional to the coulomb *energy* between the effective nuclear charge and the bonding electrons in their average position halfway between the two nuclei, which is also the distance of the covalent radius. Coulombic force is charge product divided by distance squared, and coulombic energy is charge product divided by distance. Hence the relationship, $E = CrS$. It is interesting to note here that whereas electronegativity has usually been invoked solely to explain *uneven* sharing of bonding electrons in covalent bonds between *unlike* atoms, **it also plays an equally vital, though much less appreciated, role in the even sharing characteristic of nonpolar covalence between like atoms.**

Only a very small concentration of K_2 molecules has been observed in the vapor above 760°C, the boiling point of potassium, most of the vapor being monatomic. It is not surprising that most of the weakly bonded K_2 molecules are dissociated at so high a temperature. Below 760°C the element exists as a lustrous liquid, which solidifies at 63.7°C to a typically metallic-appearing solid. We can explain this failure of the K_2 molecules to persist—studies of the solid show that they are no longer present as such—on the basis of the atomic structure of the potassium. Examination of the K_2 molecule shows us that once the single covalent bond has been formed, neither atom retains any ability to form additional covalent bonds. It also shows us that each atom has relatively low energy orbitals that are not engaged in the bonding. Under such circumstances, it is generally observed that the bonding electrons tend to become *delocalized,* abandoning their concentrated position in the internuclear region between just one pair of atoms in order to spread out among many atoms. This kind of delocalization of outer electrons into all available orbitals, minimizing the repulsions among electrons, is the distinguishing characteristic of the metallic state. Knowing this, we should then accept the familiar existence of elemental potassium as a metal, rather than a diatomic gas, as the expected behavior of atoms having the general nature exhibited by atoms of potassium.

The advantage to potassium of condensing from the diatomic gas to the metallic solid is evidenced by the atomization energy of potassium metal, which is found to be 21.4 kpm. This is to be compared with the energy to liberate one mole of potassium atoms from the diatomic molecules, which is half of 13.2, or 6.6 kpm. In other words, potassium atoms in the metallic state are subjected to cohesive forces more than three times greater than in the diatomic molecule. From the study of the structure of potassium, it is known that the atoms form the body-centered cubic lattice in which each interior atom is in direct contact with eight neighbors at the corners of an enclosing cube and about 15% farther away from the central atoms of the six adjacent cubes. Clearly the bonding forces per atom pair must be less than in the diatomic molecule, which would therefore cause the atoms to be farther apart in the solid. The metallic radius of potassium is found to be 2.35 Å, in contrast to the nonpolar covalent radius of only 1.96 Å. But potassium atoms are relatively low in atomic weight, and the metal must therefore be quite low in density. In fact, the density is only 0.86 g/ml. From this we find that the volume occupied by one mole of potassium atoms, or the atomic volume, is 45.3 ml, larger than for most of the other chemical elements and certainly most of the closely packed metals.

In summary, thus far we have seen that the potassium atom has qualities entirely consistent with its special electronic configuration and with the properties of potassium atoms joined together by covalent or metallic bonds. Although we are not yet able to predict the exact temperature of melting or boiling or the exact energies involved in phase changes, we can nevertheless appreciate that most if not all of what is known about potassium metal is consistent with what is known about potassium atoms.

Chlorine

In contrast to potassium atoms, chlorine atoms come at almost the end of their period. By this time the outermost shell has been provided with seven of the eight possible electrons, leaving but one vacancy and therefore one half-filled orbital. Whereas potassium is barely beyond the point of practically zero effective nuclear charge, chlorine represents the maximum effective nuclear charge in its period. The increase in atomic number from sodium, 11, to chlorine, 17, involves addition of outermost electrons while increasing nuclear charge, but the increased charge is not effectively blocked by the additional electrons. Chlorine being six steps beyond sodium and the effective nuclear charge having increased by about two-thirds with each step, the final effective nuclear charge of chlorine must be about 4 greater than of sodium. Therefore the atom must have a much more compact electronic cloud in chlorine than in sodium. Furthermore, the much *larger* effective nuclear charge now acts over a *shorter* distance, which **increases** the electronegativity. If chlorine atoms are smaller and more electronegative than sodium atoms, they are certainly smaller and more electronegative than potassium atoms. In fact the covalent radius, determined as half the bond length in the diatomic molecule, is 0.99 Å, only about half that of the potassium atom. The electronegativity is 4.93, much larger than in potassium.

Just as in the case of potassium, it is easy to predict that with one half-filled outer orbital per atom, chlorine will unite by single covalent bonds to form diatomic molecules. Unlike potassium, however, chlorine has lone pairs of electrons in the remaining orbitals, not vacancies. Therefore there is no possibility of Cl_2 molecules undergoing further combination, their only interattraction being weak van der Waals. The weakness of these forces prevents chlorine from condensing at ordinary conditions. It remains a diatomic molecular gas until cooled to about −34°C. At this temperature the van der Waals forces become sufficiently effective to liquify it.

We should expect the bond in Cl_2 to be much stronger than in K_2, since the effective nuclear charge in chlorine is much larger than in potassium and acts over only about half the distance. In fact, the experimental dissociation energy of Cl_2 is about 58 kpm. If we calculated this according to the $E = CrS$ relationship, E would be about 78, or 20 kpm higher. As we shall see later, both quantities are valid and have their special applications. Our present purposes are adequately cared for by the experimental value 58.

The experimental bond length in Cl_2 is 1.98 Å, making the covalent radius 0.99 Å. The combination of small radius and high electronegativity, together with the presence of the essential vacancy in the outermost principal quantum level, should give chlorine atoms both a strong hold on their own electrons and a strong attraction for an additional electron from some other atom. Consistent with these predictions are the ionization energy, 301 compared to only 102 for potassium, and the electron affinity. When a gaseous chlorine atom acquires an extra electron to become a gaseous chloride ion, about 86 kpm is released as the electron affinity.

In summary, chlorine as the free, elemental gas has properties quite consistent with the nature of chlorine atoms as we know it. The wide differences between metallic potassium and gaseous chlorine are easily rationalized in terms of the differences between potassium atoms and chlorine atoms. Potassium atoms, which are large, hold their outermost electrons only weakly and have very low electronegativity. Chlorine atoms, which are small, hold their outermost electrons very tightly and have very high electronegativity. These differences are easily interpreted on the basis of atomic structure and of its effect on how strongly the nuclear charge is evident within the outermost orbitals of the atom.

POTASSIUM CHLORIDE

KCl Gas Molecules

We have seen that potassium atoms are held together in metallic potassium with sufficient force that it takes 21.4 kpm to liberate them. Furthermore, chlorine atoms are held together in gaseous chlorine molecules with sufficient force that it takes 29.1 kpm (half of 58.2) to liberate them. It is therefore unreasonable to expect that either of these elements will exhibit any additional tendency to react unless some advantage is attainable. If we bring potassium metal and chlorine gas together, we therefore need to consider not only the bonding capacity of potassium and chlorine atoms, but also whether any combination would be advantageous enough to overcome the forces that already occupy all the atoms involved.

The bonding capacity of these elements presents no problem. We have already seen that both potassium and chlorine atoms possess one half-filled orbital per atom and thus the capacity to form one single covalent bond. If potassium atoms should combine with chlorine atoms, then, there is no difficulty about predicting that the combination should be 1:1, giving a compound of empirical formula, KCl. Whether this possibility will be realized or not, however, depends on the nature of the bond between potassium and chlorine and how it compares with the bonding in the potassium metal and the chlorine gas molecules. Let us therefore consider bringing one potassium atom and one chlorine atom together, and let us examine what would happen.

Here is where the fun and excitement begin in earnest, for if we can apply what we know of the atoms to predicting their behavior together, we have already reached a fascinating degree of understanding.

As the atoms come close enough for orbital overlap, the unpaired electron of each atom finds itself accommodated within an orbital region of the other atom that was previously vacant. It still remains, however, under the influence of the nucleus which initially claimed it as exclusive owner. Thus it is now one of a pair under the direct influence of both nuclei simultaneously. But these influences cannot be the same in both directions. In the direction of the chlorine atom's nucleus, the bonding electron pair feels a larger effective nuclear charge than in the direction of the potassium nucleus. Furthermore, the chlorine nucleus is closer at hand. What is still more important is the fact that, however we may choose to describe the condition of the bonding electrons with respect to the molecule, there is no restriction to their adjustment to the most stable average position. In other words, they are free to move. Being free to move, and being attracted toward one nucleus more strongly than toward the other, the bonding electrons will naturally be expected to adjust their average positions such that they will finally be equally attracted by both nuclei.

The bonding electrons will accordingly tend to spend more than half the time more closely associated with the chlorine nucleus than with the potassium nucleus. This must have a profound effect on both atoms, as described below:

(1) *Charge.* Since the two atoms were initially electrically neutral, the shift of bonding electrons to favor the chlorine nucleus is the equivalent of destroying the neutrality. The chlorine atom now acquires a partial (less than unit electronic) negative charge, because on the average there are now more than 17 electrons in its cloud. The potassium atom now acquires a partial positive charge, because on the average there are now fewer than 19 electrons in its cloud.

(2) *Radius.* The increase in average electronic population around the chlorine nucleus means an increase in the interelectronic repulsions and an increase in the screening of the nuclear charge. Both effects must result in expansion of the electronic cloud. The chlorine atom grows larger. Simultaneously, the decrease in average electronic population around the potassium nucleus reduces the interelectronic repulsions and also reduces the shielding of the nuclear charge. Thus the remaining electrons are encouraged to come closer together and the cloud shrinks. The potassium atom grows smaller.

(3) *Electronegativity.* In the chlorine atom the effective nuclear charge is reduced and forced to act over a longer distance, which means that the electronegativity decreases. This is exactly what we should expect, for the attraction an atom exerts on extra electrons must certainly diminish to the extent that the atom succeeds in acquiring them. But in the potassium atom the now larger effective nuclear charge is acting over a shorter distance, which means that the electronegativity increases. The reasonable assumption is made that **these adjustments cease at the point where the two electronegativities become equal** through the process of *uneven* sharing of the bonding electrons.

In order to understand how these adjustments may affect the nature of the covalent bond that exists between the potassium and the chlorine atom, we must be able to determine quantitatively the following: (1) What is the intermediate electronegativity in the KCl molecule? (2) What are the partial charges on the individual atoms that must result from this equalization of electronegativities? (3) What effect do these adjustments have on the atomic radii and therefore on the length of the covalent bond? All of these questions can be answered in a reasonably simple and satisfactory manner.

(1) The intermediate electronegativity in a compound is found to be well represented as the geometric mean of the electronegativities of all the individual atoms before combination. The geometric mean for KCl is the square root of the electronegativity product, 0.42 × 4.93, or 1.44. This is the electronegativity of potassium and of chlorine in the compound KCl.

(2) **The partial charge on a combined atom is defined as the ratio of the change in electronegativity undergone by the atom in forming the compound to the change it would have undergone in acquiring unit charge.** The latter change has been found to correspond to $2.08\sqrt{S}$, which is 1.35 for potassium and 4.62 for chlorine. In combining with an atom of chlorine, an atom of potassium has changed in electronegativity by the KCl electronegativity 1.44, minus 0.42 for potassium, or 1.02. If the atom of potassium had lost an electron completely, it would have changed by 1.35. The partial charge on the combined potassium atom in KCl is therefore 1.02/1.35 = 0.76. In combining with an atom of potassium, an atom of chlorine has changed in electronegativity by 1.44 − 4.93, or −3.49. In becoming chloride ion, the chlorine would have changed by 4.62. The partial charge on chlorine in KCl is therefore −3.49/4.62 = −0.76. The sum of the two charges is of course zero, the charge on the KCl molecule.

(3) The radius of a charged atom in a simple binary compound can be expressed as a linear function of charge: $r = r_c - B\delta$. In this equation r_c is the nonpolar covalent radius, δ the partial charge, and B an empirical factor. The radius of a potassium atom bearing a charge of 0.76 is thus found to be 1.13 Å. The radius of a chlorine atom bearing a charge of −0.76 is found to be 1.54 Å. The sum, 2.67 Å, is the calculated bond length in the KCl molecule. The experimental value is 2.67 Å.

In summary, the potassium atom, in changing to the intermediate electro-negativity in KCl, has lost 0.76 electron, leaving its average electronic population at 18.24 electrons. This has caused contraction of the potassium radius from the nonpolar 1.96 Å to 1.13 Å. In combining with the potassium atom, the chlorine atom has gained 0.76 electron, changing its average electronic population to 17.76 electrons. This has caused expansion of the chlorine atom from its nonpolar covalent radius of 0.99 Å to 1.54 Å in the KCl molecule. But by means of these adjustments, the molecule has become stabilized.

Finally we are prepared to examine the nature of the K—Cl bond more closely. Specifically, we wish to know how much energy is released when this bond is formed by the combining atoms. To calculate this, we need to analyze the polar covalent bond. The simplest procedure is to accept the familiar concept of two extreme forms of bonding: **nonpolar covalent,** in which the two electrons are evenly shared; and **ionic,** in which the two bonding electrons are completely monopolized by one of the two atoms. Any actual polar covalent bond, as would exist between any two different elements, can then be imagined as a blend, a weighted blend, of the two extreme forms. If we can calculate the energy of a nonpolar covalent single bond and also the energy of interaction between two oppositely charged ions, then we need only the correct blending coefficient to weight the relative contributions of these kinds of energy to the actual bond energy.

The nonpolar covalent energy of a heteronuclear bond, as in K—Cl, appears to be well represented as the geometric mean of the two homonuclear single covalent bond energies. For KCl this means the square root of the product 13.2 × 58.2, or 27.7 kpm. However, any change in bond length over the sum of the nonpolar covalent radii would affect this average value, since shorter bonds are stronger. For example, the sum of the nonpolar covalent radii of K and Cl is 1.96 + 0.99 = 2.95 Å, whereas the actual experimental (and calculated) bond length is only 2.67 Å. Polarity typically shortens bonds. The reason can be seen in the example of KCl when we review the effect of electronegativity equalization. Whereas the chlorine atom increased in radius by 1.54 − 0.99 = 0.55 Å, the potassium atom decreased by 1.96 − 1.13 = 0.83 Å, leading to a net contraction of 0.28 Å. But this seems certain to increase the bond strength, including the covalent contribution, so the covalent energy must be corrected by the factor, the ratio of the nonpolar covalent radius sum, R_c, to the observed bond length R_0: R_c/R_0.

For the KCl molecule, then, the nonpolar covalent energy E_c would be

$$E_c = \frac{27.7 \times 2.95}{2.67} = 30.6 \text{ kpm}$$

The ionic energy may be calculated simply as the electrostatic energy of two oppositely charged ions separated by the bond length

$$E_i = \frac{332}{2.67} = 124.3 \text{ kpm}$$

The factor 332 converts to the units kpm when the bond length is in angstrom units.

Tentatively, we have concluded up to now that the bond energy in the molecule of KCl ought to be somewhere between 30.6 and 124.3 kpm. But where? The vital importance of correctly selecting the blending coefficients may be appreciated. Fortunately, they may be determined with beautiful simplicity. The ionic blending coefficient t_i is simply the numerical average of the partial charges on the bonded atoms. The covalent blending coefficient t_c is simply equal to $1.00 - t_i$. Thus for KCl, the value of the ionic coefficient t_i is 0.76, and the covalent coefficient t_c is 0.24. The total bond energy may then be calculated as

$$E = t_c E_c + t_i E_i = 0.24 \times 30.6 + 0.76 \times 124.3 = 7.4 + 94.5 = 101.8 \text{ kpm}$$

The experimental value is 101.6 kpm, in practically perfect agreement.

Through the reasonable application of relatively very simple concepts that describe a polar covalent bond quantitatively, we have thus succeeded in accomplishing, from a few basic facts about the nature of the individual atoms, what has not been accomplished through fifty years of quantum mechanics. Since no real understanding of chemistry can ever be achieved without some quantitative knowledge of chemical bonds, we must grasp such knowledge wherever we can get our hands on it.

Now, before paying further attention to potassium chloride as a chemical substance and seeing whether its properties also conform to our expectations, let us pause to consider the blending coefficients. The average skeptical scientist may take a dim view of electronegativity values not very precisely defined, and the concept of electronegativity equalization may not prove altogether appealing. Even if the view of these scientists is sympathetic, the choice of the geometric mean for the molecular electronegativity may seem unduly arbitrary, and the calculation of partial charge may not seem acceptably satisfying. Just for a brief but enlightening moment, then, let us freely abandon all concepts of electronegativity, equalization, and partial charge, and focus our attention on determining the blending coefficients in some *entirely independent manner.*

Let us assume only that the polar covalent bond energy has been correctly expressed as the weighted sum of a nonpolar covalent contribution and an ionic contribution and that the equations for the covalent and ionic energies are correct. If so, then the ionic blending coefficient t_i can be calculated as

$$t_i = \frac{E - E_c}{E_i - E_c}$$

For KCl, this becomes

$$t_i = \frac{101.6 - 30.6}{124.3 - 30.6} = \frac{71.0}{93.7} = 0.76$$

Notice that **this is exactly the value calculated from electronegativities and partial charges.** One must therefore accept that whatever else these particular partial charges may mean, they do correctly represent the relative weights of ionic and covalent energies in the total energy, they do permit the calculation of the correct bond length, and of course they permit the accurate calculation of the energy of the polar covalent bond. Alternative assignments of atomic charges, obtained by different methods, ought to pass some similar utility test and not be automatically acceptable because of their derivation from more sophisticated concepts.

Why, when potassium atoms are already utilizing their bonding capacity in the metallic crystal and chlorine atoms are already fully engaged in diatomic molecules, should these atoms willingly break loose from their environment to join in forming KCl molecules? We have now reached the point of answering this question. It costs 21.4 kpm to isolate potassium atoms from the metal, and it costs 28.1 kpm to separate chlorine atoms from their diatomic molecules, the total being 49.5 kpm just to prepare the atoms for forming one mole of KCl. The energy released when one mole of potassium atoms joins one mole of chlorine atoms in the gas phase is, as we have just seen, 101.6 kpm. Consequently, there is a 52.1 kpm advantage in forming the KCl. In other words, the greater strength of the bond in KCl is more than adequate to supply the energy for atomizing the component elements from their standard states, and the difference is given out as the enthalpy of the reaction. This difference is, in fact, the "standard heat of formation" of KCl(g).

Changes in free energy, not enthalpy, determine the direction of spontaneous reaction, however, and thus far we have considered only the enthalpy. We may make a reasonable estimate of entropy effects by considering primarily the phase changes and possible changes in number of moles of gas, for such changes involve more significant entropy changes than mere changes from one gaseous or solid substance to another. A considerable increase in entropy is involved in atomizing the potassium, since in the vapor the atoms have maximum randomness and in the crystal are highly restricted and orderly. Overall, for the reaction,

$$2K(c) + Cl_2(g) = 2KCl(g)$$

the total number of moles of gas increases from one to two. Therefore we can predict with confidence that the entropy change will be positive, which means that

$T \Delta S$ must be negative. Since ΔH is negative, ΔG will be even more negative than ΔH. By the criterion of negative ΔG, this reaction is surely reasonable. In fact, the experimental value of the free energy of formation of KCl(g) is -56.2 kpm, 4 kpm more negative than the enthalpy, verifying our prediction.

Now we know why potassium and chlorine react to form KCl gas. It is primarily because the chemical bond in KCl is stronger than the average bonding in potassium metal and diatomic chlorine gas. But we have not yet gone to the real heart of the problem, to ask the question, WHY is the bond in KCl stronger?

To answer this more fundamental question, let us return to the separate covalent and ionic energies that were calculated, 30.6 and 124.3 kpm. It is clear that the ionic energy is greater, and that **this must always be true**. The ionic energy of a polar covalent bond replaces a part of the covalent energy but **is always greater than that which it replaces**. This is why polarity always increases bond strength. Let us consider the K—Cl bond if there were not polarity, if potassium and chlorine were initially alike in electronegativity. Then the total bond energy would be merely the 27.7 geometric mean, whereas 49.5 kpm is needed merely to prepare the potassium and chlorine atoms for combination. If it were not for the electronegativity difference between potassium and chlorine, they would not unite to form KCl gas. The governing reason for the combination of potassium with chlorine is therefore the initial electronegativity difference. It leads to a considerable ionic character in the polar covalent bond, which makes it much stronger than the bonding in the elements.

KCl as Nonmolecular Solid

We have been postponing coming to grips with the well-known fact that KCl is not actually a gas or a molecular substance except under unusual conditions. Normally it is a crystalline solid, having the rocksalt structure with each atom being surrounded octahedrally by six of the other kind. Why? This calls for a consideration of the nature of the KCl molecule.

When any atom, such as that of potassium, has one or more vacant orbitals in its outermost principal quantum level, it has the potential of acting as electron pair acceptor to an appropriate donor. In a potassium atom this potential would be very small because of the weakness of the effective nuclear charge. However, when charge has been withdrawn from any atom giving it a partial positive charge, this must enhance its potential as an electron pair acceptor in proportion to the magnitude of the charge. For example, in a KCl molecule the partial charge of 0.76 on the potassium atom makes the three unoccupied orbitals in its outer shell much more susceptible to coordination than when the atom is neutral. Similarly, when any atom has one or more electron pairs in the outermost principal quantum level that are uninvolved in the bonding, it is potentially an electron pair donor for formation of a coordination bond. This tendency would be greatly enhanced in proportion to the partial negative charge that the atom might acquire through its polar covalent bonds. For example, in a KCl molecule the chlorine atom with its three lone pairs of electrons has greatly increased potential as an electron pair donor because of its high partial negative charge of −0.76.

When two or more KCl molecules come together, not only do the oppositely charged atoms of different molecules attract one another but also **coordination can occur** between potassium as electron pair acceptor and chlorine as electron pair donor. We could not necessarily predict that the *rocksalt* structure would result from such coordination, but it is clearly a reasonable possibility considering the intermolecular forces that can be exhibited by KCl gas molecules, with each kind of atom tending to surround itself by the other kind. We can calculate the energy of such condensation to nonmolecular solid by recognizing certain distinctions between the conditions of the combined atoms in a gas molecule and in the nonmolecular solid. First, in the latter all outermost electrons may become involved in the bonding per atom pair instead of only two electrons as in the KCl molecule. Therefore we can calculate the **covalent energy** contribution as before but multiply it by four, which is the number of covalent bonds that might be formed by eight electrons per atom pair. It would make no difference whether the actual coordination were 4:4, 6:6, or 8:8—only four pairs of electrons, or the equivalent, are available.

For the **ionic energy**, we must recognize that in a nonmolecular solid any given atom must be under the influence of all the other atoms in the crystal. The net result must multiply the electrostatic attractions but also take into account that closer approach of oppositely charged atoms is restricted by the repulsions among their electronic clouds. We therefore use the familiar Born–Mayer equation. For unit charged ions, as would be formed by potassium and chlorine, we simply multiply the normal ionic energy for a polar bond by the Madelung constant, which takes into account the crystal environment, and the repulsion coefficient k, which corrects for the inter-cloud repulsions. In summary, for the nonmolecular solid form of KCl, the total atomization energy of the crystal is represented as

$$E = \frac{t_c n R_c (E_{K-K} E_{Cl-Cl})^{1/2}}{R_0} + \frac{332 M k}{R_0}$$

For potassium chloride, $n = 4$ as previously discussed, the geometric mean homonuclear single bond energy is 27.7 kpm, and R_c is 2.95 Å as before. However, since six bonds are formed with only eight electrons, each bond is about 2/3 of a normal covalent bond, and the bond length must therefore be greater than in the single molecule. We can calculate the radii of the atoms in the solid by a method analogous to that used for the gas. The radius of potassium having charge 0.76 is here 1.23 Å and that of chlorine having charge −0.76 is 1.90 Å. The sum, 3.13 Å, is the calculated bond length. It agrees well with the experimental value of 3.14 Å, which, as expected, is considerably larger than the 2.67 of the gas molecule. The Madelung constant M for the rocksalt structure is 1.75, and the repulsion coefficient for KCl is 0.90. The partial charges and therefore blending coefficients are of course unchanged from the gas molecule. We then may calculate the atomization energy of solid KCl as

$$E = \frac{0.24 \times 4 \times 27.7 \times 2.95}{3.14} + \frac{0.76 \times 332 \times 1.75 \times 0.90}{3.14}$$

$$= 25.0 + 126.6 = 151.6 \text{ kpm calculated}$$

The experimental value is 154.6 kpm, differing from the calculated value by only about 2%.

Calculation by this new **coordinated polymeric** model, rather than by using the old "ionic" model, shows that condensation of KCl gas molecules to a nonmolecular solid is favored by an increase in covalent as well as ionic energy. The energy of condensation, measured as enthalpy change, is −49.8 kpm calculated and −54.0 kpm experimental. The decrease in entropy accompanying condensation from the gas would be appreciable but not nearly enough to counteract the enthalpy change. Therefore the free energy of condensation would be predicted to be somewhat smaller than the enthalpy but still very negative. The experimental free energy of condensation is −41.4 kpm.

The formation of potassium chloride from potassium metal and chlorine gas is seen to be greatly favored not only by the electronegativity difference between potassium and chlorine, a natural consequence of their atomic structures, but also by the possibility of condensing to a nonmolecular solid, likewise the result of the atomic structures. Considering the magnitude of the cohesive forces, we should not be surprised that potassium chloride melts at 772°C and boils at 1407°C. The density is of course directly calculable from the size and weight of its atoms and the structure of its crystal. It is 1.99 g/ml. To account for its considerable solubility in water would require a study of the hydration energy of potassium and chloride ions. It is very useful to think of the solid as a coordination polymer, for then one recognizes that the process of dissolution is not so profound as might seem. In effect, it is merely a change in coordination environment. Potassium ions that were coordinated with a sphere of chloride ions in the crystal change to a coordination sphere of water molecules in the solution, and chloride ions that were surrounded by potassium ions in the crystal become surrounded by water molecules in the solution.

Chemically, potassium chloride is susceptible to strong oxidizing agents; but the negative chlorine does not easily release its surplus charge, since the chlorine was originally so high in electronegativity. The electron can be restored to the potassium atom only by an electrode containing a negative charge and when reactive chemicals are absent around the potassium ion. The chlorine in KCl, with its high negative charge of −0.76, must surely have lost its great oxidizing power, and KCl is not at all a chlorinating agent. But the negative chlorine with its lone pair electrons can act as donor in the formation of complex chlorides through coordination, such as, for example, in $KAlCl_4$. A major tendency for halides is to undergo hydrolysis. We should not expect KCl to exhibit this tendency at all, because hydrolysis depends on replacement of the halogen by hydroxide. If the hydroxide is completely dissociated, as is KOH, then there is no incentive for the hydrolysis reaction to proceed. Hydrolysis of halides occurs only if the hydroxide formed is weakly basic or acidic. In general, we find the physical and chemical properties of potassium chloride to be quite consistent with the charge distribution that we have been picturing.

THE BURNING OF CARBON

No single chemical reaction is more important from the standpoint of producing useful energy for the needs of civilized man than the burning of carbon. The energy from this reaction has probably been measured as carefully as any, and certainly there have been thousands, if not millions, of quantitative determinations of the heating value of various carbonaceous fuels. The total failure, until a few years ago, to be able to calculate the heat of combustion of carbon on theoretical grounds stands as a monument to the impotence of quantum mechanics to deal with so complex a system as a molecule of carbon dioxide. Yet, in the mid-sixties, it became apparent that this formidable assemblage of 22 electrons and three nuclei could be represented very nicely by simple calculations requiring only about fifteen minutes with a cheap slide rule.

The slide rule operator must, of course, already know something about the special qualities of carbon atoms and oxygen atoms, and he must know the experimental bond length in carbon dioxide, unless he has some way of calculating it. Let us consider these special qualities and how they affect the physical and chemical nature of the elements. Then we shall be prepared to consider the interaction of the two elements.

Carbon

A carbon atom has the electronic configuration represented as 2–4, which implies the presence of four outermost vacancies and hence four half-filled orbitals available for bonding. The atoms are small, having a nonpolar covalent radius of 0.77 Å; and since they represent the midpoint in the filling of the outermost shell from 0 to 8, they exhibit substantial but intermediate effective nuclear charge. This charge is not only sufficient to keep its own electronic cloud fairly compact. Also, being operative over such a relatively short distance, it corresponds to a fairly high electronegativity, 3.79. From the $E = CrS$ relationship, we calculate a homonuclear single covalent bond energy of 83.2 kpm, which appears to be a satisfactory average value for carbon to carbon bonds in hydrocarbons.

There should be no difficulty in predicting the joining together of carbon atoms in a symmetrical solid lattice in which each interior atom is attached by a single covalent bond to each of four neighbors at the corners of a regular tetrahedron (since this is the structure that would minimize repulsions). This form of carbon does indeed exist as diamond. Its atomization energy, 170.8 kpm, suggests a single bond energy of half that, or 85.4 kpm, since breaking two bonds per carbon atom is sufficient to atomize the diamond structure. This is a little higher than the homonuclear single bond energy given above, but this is consistent with an observed tendency for C—C bonds in hydrocarbons to become stronger the more nearly the structure approaches that of diamond.

The objection to this prediction is of course that diamond is a scarce form of carbon and a trifle less stable than the common form, graphite, for which the atomization energy is 0.45 kpm higher, or 171.3 kpm. Our knowledge of carbon atoms was evidently inadequate, for we should never have predicted the graphite structure from the qualities of the carbon atom as we perceived them. Furthermore, it hardly seems to make sense that a substantial amount of carbon liberated from its compounds is not in the form of diamond, when the energy difference from graphite is so small. However, perhaps a rationalization can be reached somewhat speculatively.

Let us suppose that we begin with unattached carbon atoms. We let them come together. The first species that can form must be C_2. The experimental evidence available about the C_2 gas species is that the two carbon atoms are held together by a double bond, as in ethylene. (This, too, needs explaining, but it seems an established fact.) We may therefore picture the C_2 species as a tetraradical, such as would be obtained by disengaging the four hydrogen atoms from a molecule of ethylene. Since the ethylene molecule is planar, we may assume that the C_2 fragment must also exhibit a planar distribution of its four half-filled orbitals. If this is so, then the most natural manner of condensation of C_2 particles together would be to form a system of planar, six-membered condensed rings exactly like graphite, since, once formed, the alternate single and double bonds would become alike in their degree of multiplicity. In other words, this speculation implies that the prevalence of graphite over diamond is largely one of mechanism of formation, and that graphite is the only way to go under normal conditions.

Regardless of the most common and abundant form of carbon, it is clear that any reaction requiring individual carbon atoms must require a high activation energy in order for the atoms to be released against the high atomization energy. Once released, carbon atoms should then be highly reactive through their moderately high electronegativity, high homonuclear single covalent bond energy, and available bonding orbitals.

We next consider the nature of oxygen atoms.

Oxygen

Being but two beyond carbon in atomic number, oxygen atoms represent systems of electronic configuration 2–6, in which the effective nuclear charge should be substantially higher than in carbon. This should cause the electronic cloud to be more compact. The nonpolar covalent radius (recently revised) is about 0.70 Å. A higher effective nuclear charge operative over a shorter distance should correspond to a higher electronegativity. Oxygen, with the value of 5.21, is more electronegative than any other element except fluorine. We can calculate by the $E = CrS$ relationship that the homonuclear single bond energy of oxygen should be 103.9 kpm.

From these properties, and in consideration of the electronic configuration which allows two half-filled orbitals per atom in the outermost principal quantum level, there should be little difficulty in predicting the physical state of elemental oxygen. It should exist as long chain polymers or perhaps in cyclic form, or both, causing it to be a **solid**. Sighing thankfully, we are glad it is not. But we need to understand why not, before we can understand why it is as it is.

We know well, of course, that elemental oxygen exists as diatomic molecules, held together apparently by double bonds, thus using the full covalent capacity of each atom in joining to the other. This fact convinces us that this form must be more stable than any single-bonded form, for there is no evidence that oxygen ever reverts to such a state. We conclude that, for oxygen atoms, one double bond is superior to two single bonds. Experimental determinations confirm this, for the dissociation energy of O_2 is 119.2 kpm, whereas a value of only about 34 kpm is obtained for the single O—O bond from a thermochemical study of hydrogen peroxide. Since this is true, then there is no problem about explaining why oxygen exists as a gas instead of a polymeric solid.

But a real understanding depends on learning *why* an oxygen double bond is so much stronger than two single bonds. This is not to be expected. A double bond places within the internuclear region twice as many electrons as a single bond and thus increases the repulsions that would tend to weaken the bond. Two single bonds would normally therefore be expected to total substantially greater energy than one double bond. In hydrocarbons that is certainly true, with twice the single bond energy, 166 kpm, being very significantly larger than the double bond energy of about 145 kpm.

This seemingly narrow little puzzle has an amazingly wide significance. A more careful consideration of the steadily changing electronic structure across a period discloses that beyond Group M4 there appears for the very first time a lone pair of electrons. Because of the limitation on vacancies, this lone pair cannot become involved in ordinary covalence. Such lone pairs occur in atoms of all the elements of Groups M5, M6, and M7. Seemingly very significantly, the occurrence of these lone pair electrons is associated with unexpected bond weakening effects. Although the explanation is only speculative, it appears that lone pair electrons may get in the way of normal bonds, perhaps by blocking off nuclear charge, so as to weaken the bonds. It is therefore assumed that the homonuclear single covalent bond energy obtained by extrapolation of the $E = CrS$ relationship is the "correct" energy, if it were completely unweakened by the lone pair electrons. (For example, this assumption allows the exact calculation of the dissociation energy of nitrogen, N_2.) It is also assumed that a triple bond, with its concentration of six electrons in the internuclear region, repels the lone pair electrons to positions wherein their weakening effect disappears. The experimental single bond energy is generally the fully weakened energy, and, in a double bond, the weakening effect is assumed to be half removed. That the weakening effect is still important in the double bond in O_2 is evident from the double bond energy of 119 compared to 145 for the carbon double bond, whereas an unweakened double bond should be stronger in oxygen.

Empirical studies of the thermochemistry of unsaturated hydrocarbons show that a triple bond is about 1.75 times stronger than a single bond of the triple bond length, and that a double bond is about 1.50 times stronger than a single bond of the same length. We can calculate what the partially weakened single bond energy in the double bond of oxygen would be by the following:

$$(119.2 \times 1.21)/(1.50 \times 1.40) = 68.7 \text{ kpm}$$

The disssociation energy of O_2 is 119.2, the multiplicity factor is 1.50, and the ratio 1.21/1.40 corrects the single bond energy to the double bond length, since 1.21 Å is the observed bond length in O_2 and 1.40 Å is twice the nonpolar covalent radius of oxygen.

By assuming 68.7 to lie halfway between the unweakened, extrapolated energy of 103.9 and the completely weakened energy, we can evaluate the latter to have the value 33.5 kpm. Thus there are three different homonuclear single bond energies for oxygen, which may be differentiated by primes: E' 33.5, E'' 68.7, and E''' 103.9 kpm.

The later chapters of this book will provide ample support for the validity of these three values. For the present let us consider only one example before attacking the carbon dioxide problem. The only example of oxygen forming a triple bond is believed to be found in the carbon monoxide molecule CO, which is isoelectronic with N_2. The geometric mean energy of unweakened carbon and oxygen is the square root of 83.2×103.9 or 93.0 kpm. The experimental bond length is 1.13 Å, and the covalent radius sum is 1.47 Å. The electronegativity in CO is the geometric mean of 3.79 for carbon and 5.21 for oxygen, or 4.44. Thus carbon has gained by $4.44 - 3.79 = 0.65$, compared to 4.05 it would have gained in becoming +1. The partial charge on carbon is therefore $0.65/4.05 = 0.16$. The electronegativity of oxygen has decreased from 5.21 to 4.44, or by 0.77, but if it had gained an electron completely, the decrease would have been 4.75. Hence the partial charge on oxygen is $-0.77/4.75 = -0.16$. The ionic weighting or blending coefficient is therefore 0.16, and the covalent coefficient is 0.84. We may now calculate the dissociation energy of carbon monoxide:

$$E = \frac{0.84 \times 1.75 \times 93.0 \times 1.47}{1.13} + \frac{0.16 \times 1.75 \times 332}{1.13} = 177.8 + 82.3 = 260 \text{ kpm}$$

The experimental value is 257 kpm, in reassuringly good agreement with the calculated value.

Carbon Dioxide

From the atomic structure of carbon, the ability to form four single covalent bonds per atom, or two double bonds, can easily be understood. From the atomic structure of oxygen, the ability to form two single covalent bonds per atom, or one double bond, can as easily be seen. There is no problem about predicting a combination having the empirical formula CO_2. Theoretically a compound of this formula might be a monomer or any multiple thereof. We know from experiment that carbon dioxide is a gas and that CO_2 is its molecular formula. The question of why it is not a polymer with single bonds will be considered presently. It is easy to account for the bonding in the molecule as consisting of two oxygen atoms joined to the same carbon atom through a double bond each. Such bonding would utilize all the covalent bonding capacity of both kinds of atoms. It would also concentrate all of the electrons around the carbon atom in two principal locations, which would therefore repel one another. The least repulsion would be realized in a linear molecule, in which the oxygen bonds are as far apart from one another as it is possible for them to be. Structural studies of CO_2 reveal that the molecule is indeed linear, with bond angle 180°.

The electronegativity difference between oxygen and carbon should cause bond polarity. The geometric mean electronegativity of two oxygen atoms and one carbon atom is the cube root of $5.21^2 \times 3.79$, or 4.69. The oxygen electronegativity has changed by $4.69 - 5.21 = -0.52$. The change corresponding to unit charge on oxygen would be 4.75. Therefore the partial charge on oxygen in carbon dioxide is $-0.52/4.75 = -0.11$. The electronegativity of carbon has increased by $4.69 - 3.79 = 0.90$. Loss of one electron would have caused an increase of 4.05. Therefore the partial charge on carbon in carbon dioxide is $0.90/4.05 = 0.22$. The average of the two charges, 0.17, is the ionic blending coefficient, leaving 0.83 for the covalent coefficient. The experimental bond length is 1.16 Å compared to the covalent radius sum of 1.47 Å.

The geometric mean of the homonuclear bond energies is here that of the carbon energy 83.2 and the E'' value for oxygen 68.7 kpm. This is the oxygen value that corresponds to a double bond, and it is still partially weakened although not nearly to the extent shown by the E' value of 33.5 kpm. The mean value is 75.6 kpm, which is only the single bond contribution. The multiplicity factor for a double bond is 1.50. We now have assembled the data needed to calculate the bond energy in carbon dioxide gas:

$$E = \frac{0.83 \times 1.50 \times 75.6 \times 1.47}{1.16} + \frac{0.17 \times 1.50 \times 332}{1.16} = 119.3 + 73.0 = 192.3 \text{ kpm}$$

Since there are two such bonds per molecule, the atomization energy is $2 \times 192.3 = 384.6$ kpm. **This happens to be exactly the experimental value.** The energy needed to atomize the carbon and oxygen is 171.3 for the carbon and 119.2 for the oxygen, totalling 290.5 kpm. The balance, $384.6 - 290.5 = 94.1$ kpm, is the standard heat of formation of carbon dioxide gas, if given a negative sign. In turn, this is identical with the heat of combustion of carbon, -94.1 kpm.

If we prefer to express this as free energy, we need to make some estimate of the entropy change that must accompany the burning. According to the chemical equation, $C(s) + O_2(g) = CO_2(g)$, the number of moles of gas remains unchanged. This fact allows us to predict that the entropy change must be quite small, and the free energy change must closely resemble the enthalpy change. Actually, the standard free energy of combustion of carbon is −94.3 kpm, almost identical with the enthalpy value.

In a pensive mood, the fireman who keeps the furnace stoked with coal might feel grateful that carbon dioxide rises through the stack and does not need to be shoveled out with the ashes. Otherwise, the question of why carbon dioxide is a gas rather than a polymeric solid seldom comes to mind. When we learn that silicon resembles carbon, it may arouse some curiosity as to why carbon dioxide is so different from sand; but like most facts that we know are so because they have always been so, we are more likely to accept them uncritically. It is now relatively easy to explain quantitatively the remarkable difference between carbon dioxide and silicon dioxide, but this explanation has two parts. First, we must explain why carbon dioxide is a gas. Second, we must explain why silicon dioxide is a polymeric solid. The latter explanation is reserved for later in this book. For the present, let us merely examine the evident preference of carbon and oxygen for double bonds over their equivalent in single bonds.

If carbon dioxide were a polymeric solid, each carbon atom would be surrounded by four oxygen atoms and each oxygen atom joined also to another carbon atom. In such a polymer, the bonding would have to be by single bonds. It is clear from the nonexistence of this polymer that one double bond is more stable than its equivalent in two single bonds in CO_2. By calculating the atomization energy of the polymer, perhaps we can find some clue as to why.

There would be no reduction of the "lone pair weakening effect" on the oxygen atoms when single bonds are formed, so the oxygen homonuclear bond energy would be 33.5 instead of 68.7 kpm. This would make the geometric mean with carbon the square root of 33.5 × 83.2, or 52.8, instead of the 75.6 used in calculating the bond energy in the monomer. Being single, the bonds would be longer, a fairly constant C—O length observed in many compounds being 1.43 Å. The blending coefficients would remain unchanged, but of course no multiplicity factor would be appropriate. We are now ready to calculate the atomization energy of the polymer:

$$E = \frac{0.83 \times 52.8 \times 1.47}{1.43} + \frac{0.17 \times 332}{1.43} = 45.0 + 39.5 = 84.5 \text{ kpm}$$

Atomization of the polymer would be equivalent to breaking four bonds per carbon atom, so the atomization energy is 4 × 84.5 = 338.0 kpm. This is 46.6 kpm less than in gaseous CO_2. And now we know why carbon dioxide does not polymerize. Beyond that, we can see why a C=O bond is stronger than two C—O bonds. In the calculation above, if we had used the E'' energy for oxygen, the covalent energy contribution would have been about 50% larger, or around 67 kpm, making the total energy about 107 or 428 for four single bonds. This energy is appreciably greater than the 384 kpm of monomeric carbon dioxide, and then carbon dioxide would indeed be a polymeric solid. The deciding factor is the **lone pair weakening of the homonuclear bond energy** of oxygen. Being much less for double bonds, this allows a double bond to be stronger than two single bonds. It appears that carbon dioxide is a gas rather than a polymer for the same reason, essentially, that causes oxygen itself to be a gas rather than a polymer.

The free energy change for the depolymerization of solid carbon dioxide would be even more negative than the enthalpy change because of the much higher entropy of the gas.

It is instructive to consider the possibility of burning carbon if there were no electronegativity difference between carbon and oxygen. If this were so, then the bond energy would be completely nonpolar covalent. Presumably the bond would also be longer, but since we do not know exactly how much longer, let us consider it unchanged, which would give a higher than realistic energy

$$E = \frac{75.6 \times 1.47 \times 1.50}{1.16} = 143.7 \text{ kpm}; \quad \times 2 \text{ bond} = 287.4 \text{ kpm}$$

Even this exaggeratedly high energy would not be enough to atomize the necessary carbon and oxygen, which would require 290.5 kpm. In other words, carbon would not burn. We could even obtain some energy from the decomposition of carbon dioxide.

Because of the characteristically large influence of bond polarity in determining bond strength, it is not surprising that **nearly all combinations of chemical elements to form compounds are exothermic** and that **compounds are greatly favored over the elements in the natural state.**

CONTRIBUTING BOND ENERGY AND BOND DISSOCIATION ENERGY

Before closing this brief preview into the possibilities of finding rational explanations for much of chemistry, let us consider one more aspect of bond energy that has been and still remains a frequent source of confusion. It has long been recognized that the average bond energy for a molecule containing more than one bond is not the same as the energy required to disengage each atom sequentially. For example, the total atomization energy of methane divided by four gives an "average bond energy" for the C—H bond, since there is no reason to suppose that these four bonds are not identical. The successive steps, $CH_4 = CH_3 + H$; $CH_3 = CH_2 + H$, $CH_2 = CH + H$; $CH = C + H$, are bound to add up to the same total atomization energy of methane, but individually the energies are all different. It has therefore become customary to refer to the energy required to break any particular bond as the **bond dissociation energy (BDE).**

This is, of course, nondistinctive for a molecule having but one bond; but where there are more than one bond, then a distinction is necessary between the BDE and that quantity of energy which the bond may be considered to contribute to the total atomization energy of the molecule. Since the latter can only be determined experimentally as an average bond energy when the bonds of the molecule are all alike, it seems more appropriate to call this energy the **contributing bond energy (CBE)**. For instance, the atomization energy of water (gaseous) is about 222 kpm, from which we conclude that each bond contributes an equal amount, 111 kpm. This value, 111, is the contributing bond energy (CBE). But dissociation of water to H + OH requires about 120 kpm, the BDE of this bond. The dissociation of the second hydrogen, OH = O + H, then has a BDE of 102 kpm. Whatever the mechanism, complete atomization of water requires the same total energy, 222 kpm, but the actual energy for dissociating either bond separately is not the same as the average bond energy.

The bond dissociation energy may be greater or less than the contributing bond energy, as in fact just illustrated. Unfortunately, we do not yet understand enough about bonding to predict or calculate bond dissociation energies in all cases. However, they can be extremely important in their influence on the nature of chemical reaction, and they can often be evaluated by experimental methods. In contrast, contributing bond energies are accessible to experimental determination only when all the bonds of a molecule are alike. When the bonds are not all alike, then the individual values can only be determined by the methods previously described herein for KCl and CO_2. Consequently there is then no direct check on their accuracy. We can only assume, reasonably, that if the sum of the calculated contributing bond energies for the several bonds of a molecule is equal to the experimental atomization energy, the individual contributing bond energies are probably reliable.

Carbon dioxide happens to present an example for which the bond dissociation energy can be treated quantitatively. It has been found experimentally that one oxygen atom can be broken loose from a carbon dioxide molecule at the cost of about 127 kpm. This might initially seem strangely low, considering that the contributing bond energy is about 192 kpm. Why is breaking this bond 65 kpm easier than we should think it should be?

To answer this question, we need to examine what happens to the remaining fragment after we have broken off one oxygen atom. Whatever readjustment may be undergone by a single atom breaking loose is already taken fully into account by the normal bond energy concepts. But a polyatomic fragment may undergo a readjustment or reorganization that changes the nature of the remaining bonds. In this example, what is left is the remnant, CO, which can exist by itself as a stable molecule. The strengthening of this bond that takes place simultaneously with the breaking of the other bond to oxygen provides energy that helps in the breaking of the other bond. The remaining C—O bond strength increases from its original 192 to 257, or by 65 kpm. Therefore the energy needed to break the other bond is reduced from 192 to $192 - 65 = 127$ kpm.

Whenever a molecule consists of more than one bond, dissociation of one bond will create at least one polyatomic fragment and possibly two. If either fragment can easily reorganize to form a stable molecule of its own, then it is easy to determine the reorganizational energy (E_R) by subtracting the sum of the contributing bond energies of the fragment before dissociation occurs from the atomization energy of the new molecule. However, if the radical released cannot form a stable molecule, then we may not be able to explain the reorganizational energy even though we may be able to calculate it.

Additional features of carbon dioxide and its chemistry will be discussed in later chapters.

SUMMARY AND COMMENTS

The two examples discussed herein were taken from hundreds of possibilities, with the intention that they be fairly representative of what is available and what is possible. I believe they are sufficient to allow a reasonably clear view of the very exciting potential for understanding and explaining chemistry that now exists. As a teacher, I have been trying for many years to bring this excitement to the classroom. It would be dishonest to pretend that the fascination has been universally infectious. On the other hand, this approach does indeed offer the opportunity for a greater depth of understanding of common chemical phenomena than has heretofore been available. The quantitative determination of the energies of polar covalent bonds is a giant step forward, but it is only fair to call attention to the obvious—that once individual substances can be accounted for, chemical changes also come within simple comprehension. What is closer to the heart of chemistry?

It is too much to expect that universal agreement could ever be reached on the very important matter of what exactly is meant by "understanding." To me, all human understanding relative to the absolute in wisdom and comprehension is to some extent illusory. But this does not mean that it is undesirable or lacks worth. The illusion of understanding, when it permits logical and reasonably correct predictions of the nature of compounds from a relatively simple knowledge of the nature of their atoms, is perhaps about as close to achieving a real understanding of chemistry as we mere humans may hope for. To the great majority of practicing chemists and related scientists, I believe this illusion can be helpful toward filling a real need.

Finally, one may dismiss the disclosures of this chapter as though they were lucky accidents of the black arts and ask, "Just exactly what can be learned by this approach that is both new and valid?"

It is appropriate here to take space only for a brief and partial list.

(1) A quantitative relationship among three basic properties of an atom: electronegativity, nonpolar covalent radius, and homonuclear single covalent bond energy. The availability of the latter two quantities from experiment permits a much improved evaluation of the electronegativity of many of the chemical elements.

(2) A quantitative means of evaluating the partial atomic charges that result from initial differences in the electronegativity of atoms that form a compound.

(3) For binary compounds, an accurate method of calculating bond length as the sum of the radii of the two atoms, adjusted for partial charges.

(4) A quantitative description of polar covalence. A polar covalent bond can be treated as a blend of nonpolar covalence and ionicity and its energy as the sum of separately calculated contributions, partial charges being indispensable in determining the blending coefficients.

(5) Multiplicity factors that permit converting single bond energy to the bond energy of a multiple bond.

(6) The first recognition of a previously misinterpreted and frequently exhibited bond weakening effect in elements of Groups M5, M6, and M7 as being an intra-atomic effect and not an interatomic effect.

(7) Recognition of the reduction of bond weakening not only by multiplicity but also in certain types of single covalent bonds.

(8) Through bond energy calculations, the estimation of enthalpies of chemical reactions and thus a fundamental understanding of chemical change.

(9) A clearer understanding of the interrelationships among bond dissociation energies, contributing bond energies, and reorganizational energies, and new methods of obtaining previously unavailable data about them.

(10) Clear, simple, fundamental, and quantitative explanations of many chemical phenomena that have previously been puzzling. Some samples are: the acid—base properties of oxides; the abnormally low single bond energies of nitrogen, oxygen, and fluorine; the striking differences between oxides of nitrogen and of other M5 elements; the contrast in properties between carbon oxides and silicon oxides; the nonexistence of sulfur iodides; the instability of SO and S_2O; the periodicity of properties of binary compounds of metal and nonmetal; the instability of carbonic acid; the nonpolymeric nature of carbon disulfide; the nonexistence of carbon—oxygen analogs of the silicones; unexpectedly high bond energies in certain binary halides; and many, many more.

None of this information, to my knowledge, has been available except through the approach detailed in this book. Nothing is deliberately concealed herein, and you will find many new questions as well as answers. This approach has been developing over a period of 25 years, the development having been almost continuously harassed and impeded by unrelenting censorship. The time for objective appraisal seems long overdue.

TWO

Significant Atomic Properties of the Chemical Elements

ATOMIC STRUCTURE

From a combination of chemical intuition, atomic spectroscopy, and wave mechanics, there has evolved a "picture" of an atom which, for most of the chemical elements, is today not seriously questioned. This is the nuclear model of Rutherford, elaborated by the concepts of electronic configuration. A tiny concentration of all the protons and neutrons of the atom, called the nucleus, is surrounded by a very thin cloud of electrons, 10^5 greater in diameter and in sufficient number to balance the nuclear charge exactly. It is the structure of this cloud that is called the electronic configuration of the atom. Electrons are arranged in successively higher energy levels around the nucleus, each level being quantized with respect to energy and limited with respect to electron capacity. Each kind of atom differs in atomic number from every other kind of atom, and thus each chemical element is distinguished by its unique atomic number.

If one lacked knowledge of the electronic configurations of the atoms, one might very well assume that feeding electrons one by one into the positive electrical field of an atomic nucleus would cause each to enter the nucleus and there combine with a proton forming, presumably, a neutron. From observation we know that this does not happen. We know that a neutron, isolated from other particles, tends to be highly unstable, decomposing to a proton and an electron and a relatively large amount of energy, with a half-life of about fourteen minutes. For this reason alone, we do not expect an electron freely to join a proton to form a neutron, despite the strong coulombic attraction between the opposite charges. Furthermore, we know enough about nuclear energy to realize that destruction of a proton by absorption of an electron would also upset the very strong interactions among the nucleons which allow the close proximity of like-charged protons despite their mutual repulsions.

One might then imagine an alternative arrangement of positive and negative charge that would minimize the energy without requiring actual coalescence of the charges. The electron(s) could move rapidly around the nucleus instead, thus creating the same effect that the center of negative charge coincides with the center of positive charge without the physical necessity of such coincidence. The example of a hydrogen atom, in which an electron is much more stable traveling around the proton than it would be joined with it to form a neutron, illustrates the practicality of this kind of electron–nucleus association. But then, one might very reasonably suppose that the arrival of each successive electron, up to the number required to balance the nuclear charge, would disrupt the interactions among the nucleus and the earlier-arrived electrons, until all electrons would occupy equivalent positions around the nucleus. This would be a physical impossibility, however, because of steric requirements and the need to minimize repulsions while maximizing attractions. In arranging to a structure of minimum energy, the electrons form in concentric levels, or energy shells, farther and farther from the nucleus. Thus it is that **each successive electron finds its position predetermined by the electrons that have come first.**

Each electron within a given atom is most likely to be found in a particular region, called an orbital. This orbital is designated by three quantum numbers. The principal quantum number indicates relatively how far removed from the nucleus this orbital is, the orbital quantum number indicates its general shape, and the orbital magnetic quantum number specifies which orbital, if there are more than one of the same principal level and shape. Finally, we have the restriction that one particular orbital can accommodate only two electrons at the same time, and then only if the electron magnets are opposed. This gives rise to the spin magnetic quantum number, bringing to four the number of quantum numbers that then exactly specify any given electron of the cloud. The Pauli exclusion principle assures us that no two electrons in the same atom can be accurately described in terms of the identical four quantum numbers. Consequently, if two electrons have the same principal quantum number and the same orbital quantum number and the same orbital magnetic quantum number, they must then occupy the identical region, or orbital. This is possible only if their spins are opposed.

From the combination of viewpoints first mentioned, it has been determined that the successive principal quantum levels that exist around an atomic nucleus are of differing capacity depending on their distance from the nucleus. In the first level, closest to the nucleus, there is only room for one orbital, the s orbital, which is spherically symmetrical with respect to the nucleus. In the second level, next farther out, there is also an s orbital, which here is supplemented by three p orbitals, roughly dumbbell shaped and cylindrically symmetrical along the X, Y, and Z axes. In the third level, these four orbitals are the most stable, and then there are five additional d orbitals which as a set are spherically symmetrical with respect to the nucleus but individually have shapes roughly described as crossed dumbbells. In the fourth, and all higher levels, there are these same nine orbitals, with an additional group of seven f orbitals of even more complex shapes. Thus we have electron capacities for the principal quantum levels 1, 2, 3, and 4 of 2, 8, 18, and 32, which can be expressed as $2n^2$.

Although the principal quantum number indicates approximately the order of magnitude of energy of the electron, an important overlap occurs such that the s orbital of $n + 1$ is always filled before the d orbitals of the n shell in the sequential building up of the chemical elements. It is this fact that determines the nature of the periodicity of the elements. As soon as a given principal quantum level is filled to 8 electrons, then the ninth electron always begins the filling of the next higher principal level instead of continuing to 9 in the originally outermost level. Consequently the buildup of the elements corresponds to the successive building up of outermost principal quantum levels to 8 electrons. This filling is interrupted by the insertion of d electrons (transitional elements) and of f electrons (inner transitional elements) at appropriate positions. In this book our primary concern is with the major group elements. This is certainly not for lack of interest, but because of complications in the treatment of transitional elements that have not yet been solved.

The principal feature of atomic structure that concerns us, in our enforced concentration on major group elements, is therefore the content of the outermost principal quantum level. We shall consider atoms that have from one to eight outermost electrons, and we shall consider the properties that result from this electronic configuration.

First, however, there is an interesting area of supplementary evidence that has not commonly been brought to appropriate attention. The great majority of chemists are in general content to know the electronic configurations of the chemical elements without truly understanding them, having accepted them on faith. Experimental evidence exists, however, relatively easily understood, which supports the conclusions of quantum mechanics and spectroscopy about atomic structure. For a considerable number of the chemical elements, it has been possible to determine the successive ionization energies all the way down to the bare nucleus. For additional values, where needed, it has been possible to make reasonable estimates through extrapolation procedures. In this manner, complete data have been assembled for all the elements from hydrogen through xenon, number 54. The least reliable estimates are probably those for the $3d$ and $4d$ electrons, but there is reason to believe them to be at least approximately correct. We can then construct an "energy profile" of the atom, in which the number of the ionization energy is plotted against the ionization energy.[1] The example of krypton is given in Fig. 2-1.

[1] R. T. Sanderson, *Chemistry* **46**, No. 5, 12 (1973).

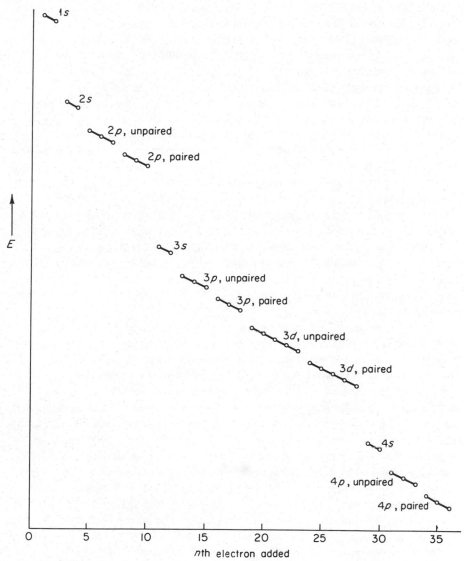

Fig. 2-1 Energy profile of a krypton atom. (The energy scale is arbitrary and greatly compressed.)

Even though for practical reasons the energy scale has been arbitrarily compressed, there should be no problem in recognizing the existence of major energy levels for electrons, corresponding to the principal quantum levels. Also one does not find uniformly changing energies within a given principal quantum level. For example, in the figure, the first three ionization energies are shown to increase in a regular pattern, but the fourth is higher than set by the initial trend. The fifth and sixth follow the new pattern, suggesting that the outermost principal quantum level contains two sets of three electrons. Then there is a larger jump to the seventh electron and a lesser one to the eighth, corresponding to a more stable pair of electrons in addition to the first six. The existence of a very large energy increase in the jump to the ninth electron corresponds to beginning the disruption of the next lower principal quantum level. A similar pattern appears, only this time there are two groups of five electrons before the two groups of three are recognizable, giving a total of 18 electrons instead of 8 in this principal quantum level.

Although it is doubtful whether anyone ignorant of other features of electronic configuration could predict from the energy profile alone how the individual principal quantum levels are structured, the profile does correspond very beautifully to the structure as determined by other means. Thus the first three electrons removed are paired electrons, one from each of three p orbitals. The reason for the jump in energy for the fourth electron is that now the remaining electrons in the p orbitals are unpaired and not subject to the repulsion of the second electron. Their removal through ionization is not, in other words, assisted by the repulsion of the other electron, as is true for the first three electrons to be removed. So, the three p orbitals are emptied out by first removing one electron from each orbital and then removing the second.

The larger energy gap to the s orbital is of course consistent with the greater stability of the s orbital compared to p orbitals of the same principal level. The very large gap to the next lower principal quantum level demonstrates that although the impression of an overlap is given by the manner of filling orbitals during the building up of elements of successively higher atomic number, within any given atom a given principal level is always filled before the next level is started. Then, the two groups of five electron energies represent the removal of paired and unpaired electrons as before. In summary, **the energy profile prepared from the successive ionization energies of an atom confirms what we have accepted largely on faith about the electronic configurations of the chemical elements.**

There is yet one feature of atomic structure that is so obvious that it is grossly neglected in most discussions. I refer to the fact that whenever there are fewer than eight outermost electrons, vacancies are at least as important as electrons. We may define a vacancy as an electron region represented by four quantum numbers but having no corresponding electron. It is therefore a region which might accommodate an electron. It is therefore a region in which the nuclear charge can be felt, for no electron can be stably accommodated by any atom unless that electron can sense, despite the repulsions of all the other electrons, a net attraction from the nucleus. Conversely, no electron can sense an appreciable net attraction from the nucleus unless it is able to occupy a region that is permissible according to quantum theory. It will be very helpful, consequently, to visualize the outermost shell of any atom not merely in terms of the number of electrons in it, but also in terms of the number of electrons which *might* be in it. The total electron capacity of the outermost shell is eight, and consequently the number of vacancies must be eight minus the number of electrons.

In summary, all atoms consist of a highly condensed aggregation of protons and neutrons, the nucleus, surrounded by sufficient electrons to balance the positive charge of the nucleus. The outermost energy level of the electron cloud can contain eight electrons (two for the first two elements). Fewer than that must signify vacancies that are capable of accommodating electrons. When atoms come near to one another, the closest contact is made by the outermost electrons. Therefore the behavior of the atoms when in close proximity must depend largely on the nature of their outermost principal quantum levels, which can be translated into number of outermost electrons and vacancies.

ATOMIC PROPERTIES

Valence

The principal information available from a knowledge of the electronic configuration of an atom is its combining capacity, known loosely as its valence. From the familiar example of the H_2 molecule, we recognize the essentials for the formation of a covalent bond: each atom must supply one outermost half-filled orbital (one unpaired electron and one vacancy). The normal covalence of an atom of the major group elements can therefore be determined on the basis of the number of outermost half-filled orbitals that are possible. It is unnecessary for the ground state to hold the outermost electrons in position for maximum bonding. For bonding, the electrons will spread out one to an orbital as far as is possible. Therefore an atom normally will form as many covalent bonds as it can supply half-filled orbitals. With up to four outermost electrons, the number of bonds is limited by the number of electrons, since there are sufficient vacant orbitals to allow up to four bonds. Beyond four outermost electrons, with five or more, the number of covalent bonds possible is restricted by the number of vacancies, which goes from three to zero, for here there are plenty of electrons but not enough vacancies.

In addition to these more conventional possibilities, we recognize examples that are believed to involve the following:

(1) The s electrons may remain paired rather than undergo promotion of one to an otherwise vacant orbital. This appears to happen chiefly when the element is fairly high in atomic number. It causes heavier elements in Group M2' to be reluctant to form bonds (Hg), in M3 to form only one bond instead of the expected three (Tl), and M4 (Pb) to form only two bonds instead of the expected four. This is called the "inert pair effect." Its exact cause is somewhat controversial. However, it appears to be assisted to some extent by the action of the lone pair electrons in

blocking off the nuclear charge so that the electronegativity is substantially lower in the lower oxidation states. This causes the bonds to nonmetals to be much more polar, and therefore stronger, than they otherwise would be.

(2) One of a pair of electrons may be promoted to an outer d orbital, thus creating simultaneously two new covalent bonding orbitals beyond the normal supply. This can only happen, of course, when an outermost lone pair of electrons exists, which means in an element of M5, M6, or M7 (possibly also M8). It does not normally happen unless the other components of the molecule were initially highly electronegative, creating a fairly high partial positive charge on the central atom which then allows the outer d orbitals to sense the nuclear charge more strongly. But the three fluorine atoms in PF_3, for example, withdraw sufficient charge from the phosphorus atom to activate outer d orbitals, and one of the lone pair electrons can be promoted to a d orbital allowing, with fluorine, PF_5.

(3) A pair of electrons in the outermost principal quantum level may interact with two other atoms, each with one outer half-filled orbital, forming a three-center bond. In molecular orbital terminology, this involves three molecular orbitals, one bonding, one nonbonding, and one antibonding. The four electrons fill the first two and leave the antibonding orbital vacant. Thus two atoms are attached to the central atom through what is equivalent to one-half bond each, since two bonding electrons are serving two bonds. This kind of bonding appears to be present in the xenon fluorides and in the halogen polyhalides and possibly in more common species such as the bifluoride ion, FHF^-, and PCl_5.

(4) A vacant outermost orbital, under appropriate conditions, may accept a lone pair of electrons from another atom, forming a coordinate covalent bond between the two atoms. This happens most commonly when both atoms have already formed their quota of normal covalent bonds.

(5) Similarly, a molecule in which an outermost electron pair is left uninvolved in the bonding may serve as an electron pair donor in coordination with the vacant orbital of another molecule.

(6) Various degrees of covalent bond multiplicity are possible that involve the sharing of more than two electrons between the same two atoms.

(7) Compounds are common in which the bonding structure cannot be represented by a single conventional assignment of bonding electrons. In most of such examples, the bonds appear to be intermediate between conventional forms. This phenomenon is called "resonance" and the limiting or alternative structures are called "resonance structures." In general, wherever sufficient electrons and/or vacancies are not available to form equal bonds to each of two or more like atoms which should otherwise be equivalent, the bonds average out alike rather than remain different. For example, the bonds in sulfur dioxide appear to be identical, but they must be described as an average of an ordinary double bond and a single coordination bond.

(8) Wherever the simplest molecule that might form still contains lone pair electrons in the outer levels and also outer vacancies, there is a strong tendency for the molecules to condense further, usually to a nonmolecular solid in which full

advantage is taken of all possible electrons and vacancies. The only known exceptions are the boron halides. In such nonmolecular solids, all of the outermost electrons, not merely those usually regarded as valence electrons, may become involved in the bonding. Such condensation is usually referred to as increasing the coordination number but not as changing the "valence."

Effective Nuclear Charge

The electronic configuration alone tells us very little else about an atom except the valence, as discussed above. What is much more informative, in a general sense, is the *effect* which that electronic configuration must have on the residual forces at the surface of an atom. This is of course because all atoms are assemblages of negative and positive charges, which are bound to interact with one another. The concept of the neutral atom is a valid one only at distances of many atomic diameters. It should be clear that when an electronic cloud surrounds a nucleus, at any given point on the surface of the atom not all the electrons can be between the nucleus and that point. Some of them must be on the far side. Even though the total negative charge is sufficient to neutralize all of the nuclear positive charge, at any given point on the surface of the atom some of the positive field of the nucleus may be sensed.

One way of visualizing the situation would be to consider that some foreign electron, approaching to the periphery of an atom, interacts attractively with the entire positive charge of the nucleus and is at the same time repelled by each of the electrons already present so that a net force may be considered to exist between the foreign electron and the atom. An equivalent alternative view is that the foreign electron only senses a small fraction of the nuclear charge, the bulk of it having been shielded or blocked off by the other electrons already present. It has been possible to assign screening constants to each electron of an atom such that the residual positive charge felt at the periphery of the atom, called the "effective nuclear charge," equals the total nuclear charge minus the sum of the screening constants. The exact evaluation of the screening constants is unfortunately impossible. It would require far more complex calculations than are attainable. But useful values have been estimated long ago by John Slater[2] and later modified by a much more sophisticated but possibly no more accurate method by Clementi and Raimondi.[3]

There is no need to concern ourselves here with the numerical values except to note one very significant point. This is the apparent fact that outermost electrons are much less efficient at screening nuclear charge than are the underlying electrons. Whereas the underlying electrons are fairly efficient, the outermost electrons only screen about one-third their maximum. This means that **when the atomic number of an atom is increased by one, by adding one more proton** (and

[2] J. C. Slater, *Phys. Rev.* **36**, 57 (1930).
[3] E. Clementi and D. L. Raimondi, *J. Chem. Phys.* **38**, 2686 (1963).

appropriate neutrons if necessary) **to the nucleus and one more electron to the outermost principal quantum level, the additional unit positive charge is only screened one-third.** Therefore the effective nuclear charge **increases by two-thirds.** For example, the effective nuclear charge acting on the outermost electron of an atom of sodium is only about +2, since the underlying 10 electrons block off most of the nuclear charge of +11 quite efficiently. The change from sodium to magnesium involves increasing the total nuclear charge to +12 and the total number of electrons to 12, but the twelfth electron only blocks off about one-third of the added proton. It is too busy avoiding the outer electron already there to spend much time between it and the nucleus. Thus the effective nuclear charge rises from 2 to $2\frac{2}{3}$. The change to aluminum causes a similar rise for similar reasons to $3\frac{1}{3}$. The effective nuclear charge is then about 4 for silicon, $4\frac{2}{3}$ for phosphorus, $5\frac{1}{3}$ for sulfur, and 6 for chlorine.

Throughout the major groups of the periodic table, the effective nuclear charge is very low at the beginnings of the periods, where the outermost shell contains only one electron, and steadily increases toward the right, becoming highest at the halogen. It is important to note here that in order for the nuclear charge to be effective to a significant degree, **there must be a vacancy capable of accommodating an electron from somewhere else.** When the last vacancy has been filled in the M8 elements, then the blocking of the nuclear charge is essentially complete in that no foreign electron can now sense an appreciable fraction of it.

The behavior of two atoms when they come together, which is at the very heart of all chemistry, is strongly dependent on whether there are outermost vacancies and **how strongly the nuclear charge can be felt within those vacancies.** Therefore the effective nuclear charge is possibly the **single most important quality or property of an atom.**

None of us wishes to indulge in circular reasoning, risking the possibility that our conclusions may be only the inevitable consequence of our conscious choice of presuppositions and represent no new progress toward revealing the truth. Up to a point this attitude represents wisdom, but I believe it can be carried too far. After all, the qualities of any atom must in one way or another reflect its composition and structure. Indeed, it would be disturbing if these properties were not fundamentally in close relationship with one another. Rather than view with apprehension such correlative success as may be attained, therefore, it would seem more appropriate to enjoy some delight whenever things come out as they are supposed to. We shall see that other important qualities of an atom are closely related to the effective nuclear charge.

Nonpolar Covalent Radius

The probability of finding the outermost electrons of an atom does not drop off suddenly when the "periphery" of the atom is reached but gradually diminishes in such a way that the radius of the atom is not thereby defined. However, two kinds of approach of two atoms can be distinguished and used to define radii. When the approach does not lead to the formation of a chemical bond and the two atoms are alike, half the internuclear distance when they are in contact can be defined as the "van der Waals" radius. This has some useful application but may be somewhat variable depending on the circumstances. When the approach leads to the formation of a chemical bond between the two atoms, then the atoms remain much closer. If the two atoms are alike and have no electrical charge, half the internuclear distance is called the "nonpolar covalent radius." It is usually not more than two-thirds as large as the van der Waals radius, but it has the advantage that it is a constant value for a given type of bond between uncharged atoms.

If two different atoms happen to form a nonpolar or nearly nonpolar bond, then the nonpolar covalent radii are such that their sum will equal the bond length. Covalent radii tend to change with charge, however, which is why the term must be modified by the word "nonpolar" if it is to have fixed significance. Also, the nonpolar covalent radius normally refers to a single bond. Bonds using more than two electrons are invariably shorter and thus the "bonding radii" must be smaller.

Since the nonpolar covalent radius is a sort of relative measure of the "size" of an atom, it must represent a dynamic balance among the various electrostatic interactions within the atom. That is, the electronic cloud of an atom tends to become as compact as possible, because of the attraction for the nucleus, and consistent with the interelectronic repulsions. We have previously considered the effective nuclear charge as that part of the nuclear positive charge that can be sensed by the outermost electrons of an atom. The effective nuclear charge must have a large influence on the size of the atom in that, if it is large, electrons will be held relatively close to the nucleus despite their mutual repulsions, but if the effective nuclear charge is small, then the electrons will spread into a greater volume and be held much less compactly.

The inefficiency of outermost electrons in blocking off nuclear positive charge from one another causes, as we have seen, a steady increase in effective nuclear charge across a period of the major group elements as the number of outermost electrons is increased from one to seven. Consequently, as the nonpolar covalent radius of the atom decreases, the atomic clouds become increasingly compact from left to right across the periodic table. In fact, the average volume per electron is much smaller in the halogens than in the alkali metals. Initiation of a new principal quantum level, however, places an electron in an outermost orbital wherein the nuclear charge is sensed only relatively faintly, so that the radius increases abruptly from halogen to the next higher alkali metal. This is the cause of the well-known periodicity of atomic radius and, in fact, the basis of the famous atomic volume curve first pointed out by Lothar Meyer.

Fortunately, many nonpolar covalent radii can be determined experimentally, some with a very high degree of accuracy. Diatomic molecules, held together by single covalent bonds, are formed by all of the alkali metals and by all of the halogens, thus providing information on the nonpolar covalent radii of these elements as half the experimental bond length. Elements that form only metallic bonds with like atoms present a greater problem, requiring estimates of nonpolar covalent radii to be made on the basis of covalent bond lengths in various compounds, preferably in which the polarity is relatively slight. But the nonmetals, with the exception of the first member of some of the groups, tend to join atoms together in a network of single covalent bonds, permitting the estimation of nonpolar single covalent bond radii as half the internuclear distance in solid crystals. All in all, we are reasonably certain of the correct nonpolar covalent radii of most of the major group elements. The data are summarized in Table 5-2 and discussed individually where appropriate under the later summary of each individual element in Chapter Three.

The nonpolar covalent radius of an atom is significant not only in its contribution to the geometry of joined atoms. It is important also through its influence on the magnitude of the nuclear force at the periphery of the atom, affecting both the electronegativity and the homonuclear single covalent bond energy.

Electronegativity

The suggestion was made by Walter Gordy and later developed more fully by Allred and Rochow[4] that the force called electronegativity is really the net attractive force experienced by an outermost electron interacting with the nucleus. As such, the electronegativity is proportional to the coulombic force between an outermost electron and the effective nuclear charge.

Pauling Electronegativities. Electronegativity was originally proposed by Pauling, as a parameter that would allow prediction of the approximate polarity of a covalent bond. The idea is that bonding electrons are not usually shared evenly by the two bonded atoms except where the two bonded atoms are of the same kind. Any bond between two different kinds of atoms is likely to be polar, because the bonding electrons initially sense different attractions toward the two nuclei. They therefore tend to spend more than half the time more closely associated with the nucleus of the atom that initially attracted them more. This imparts a partial negative charge on that atom and leaves a corresponding partial positive charge on the other atom. The atom having initially higher electronegativity is the one that becomes negative, leaving the other atom positive. The extent to which this unevenness of sharing of the bonding electrons may be expected corresponds to the magnitude of the electronegativity difference originally exhibited by the two separate atoms. This indeed has been the major application of electronegativity— toward predicting the direction and extent of polarity in covalent bonding in a purely qualitative manner.

Pauling[5] sought to evaluate electronegativity in the following way. He assumed that when two different atoms formed a covalent bond, if the electrons were

---

[4] A. L. Allred and E. G. Rochow, *J. Inorg. Nucl. Chem.* 5, 264, 269 (1958).
[5] L. Pauling, "Nature of the Chemical Bond," Third Ed., Cornell Univ. Press, Ithaca, New York, 1960.

evenly shared between the two atoms, it would be as if each atom contributed to the bond exactly as it would to a bond with a like atom. The energy of that bond would therefore be merely an average of the homonuclear single covalent bond energies of the two atoms. He expressed a preference for the geometric mean but used the arithmetic mean for greater convenience. He also observed that every actual bond between unlike atoms, every heteronuclear bond, seemed to be stronger than the average of the homonuclear energies. This "excessive" strength of the bond is released as heat. Pauling ascribed this to an "extra ionic energy" resulting from an initial electronegativity difference. In other words, he proposed that the heat of reaction, such as $H_2 + Cl_2 = 2$ HCl, should be zero except for the electronegativity difference, which makes the reaction exothermic. He evaluated electronegativities by assuming the "extra ionic energy" to be a simple function of the square of the electronegativity difference.

As we shall see, electronegativity values so determined cannot be sufficiently correct for quantitative application because of a fallacy in the basic premise. Such values have nevertheless become the most widely accepted throughout the world. They are tabulated, if little used, in practically every textbook of chemistry. They have been quite satisfactory when applied, as intended, for indicating *qualitatively* the direction and extent of bond polarity. Many alternative methods of evaluating electronegativity empirically from other properties have been proposed. Most of these have been converted to the arbitrary units of Pauling, with astonishingly good agreement with the original Pauling values.

Where, surprisingly, the Pauling electronegativities are most obviously unsatisfactory is in the reverse procedure of calculating reaction heats from electronegativity differences. Reasonably good results are obtainable when the bonds are only slightly polar, but for highly polar bonds the discrepancies are intolerable. The problem appears to arise in the basic idea of a polar covalent bond. The implication of Pauling's treatment is that the two bonding electrons of a heteronuclear bond can be shared exactly equally, full time, giving the calculated average homonuclear bond energy, and at the same time, or part of the same time, be monopolized by one of the atoms to the total deprivation of the other, forming a negative and a positive ion which then attract one another. In this sense, the "ionic energy" could be extra energy. However, it should be clear that two electrons cannot simultaneously be (1) evenly shared by both atoms and (2) monopolized by one of the atoms. In other words, the covalence must be *reduced* from its maximum if ionicity is to become involved at all. It does not make sense to picture the bonding electrons as evenly shared *all* of the time but unevenly shared *part* of that time. Rather, it is reasonable to reduce the fraction of time in which the electrons are evenly shared to allow some time when they are unevenly shared. **Ionic energy is then not a supplement to the total covalent energy, but it substitutes for part of the covalent energy.** The reason this increases the total energy, as will be shown in more detail later, is that the ionic energy always exceeds that quantity of covalent energy which it replaces.

It will become amply evident in the pages to follow that this modification of Pauling's concept of polar covalence is both essential and correct. For the present, let us examine the idea of electronegativity more closely and see exactly how the numerical values used in this book—on a scale different from the Pauling scale— were obtained. The whole concept of electronegativity has suffered in acceptance and application over the years by the lack of a precise definition permitting either exact calculations or accurate experimental measurement. Unfortunately, it is not yet possible to establish these extremely important values on an absolute, first-principle basis. Nevertheless, they have proven remarkably reliable in performing calculations of bond energies and bond lengths with much greater accuracy and wider applicability than heretofore possible.

If we could know exactly the correct screening constants of the electrons so that the effective nuclear charge could be evaluated perfectly, then, in my opinion, there would be no method to rival the Allred and Rochow method of determining electronegativity. There would still be what seem to me rather unimportant problems of which orbitals are involved in the bonding and to what extent is there hybridization and how does this affect the electronegativity. The reason I think these are not very important is that I find in extensive studies of bond energies very little evidence that a satisfactory electronegativity value cannot be assigned to an atom without allowances or adjustments for various conditions of the bonding. Further discussion of this point will be presented later.

Unfortunately, the calculation of screening constants or of effective nuclear charge from first principles would be as impossible as most calculations that deal with the interactions of many particles. But the concept of electronegativity being **proportional to the coulombic attraction between an atomic nucleus and an electron occupying an outermost orbital** is certainly qualitatively sound. As a basis for visualization, it is very satisfactory. We can easily picture an electron approaching the outermost vacancy of an atom and recognize that it must be repelled by all the electrons already there but attracted by the nucleus. If there is a net attraction, then the repulsions must be exceeded by the nuclear attraction. In fact, **the significance of an outermost vacancy is that here is a region within which an electron can sense a greater attraction than repulsion.** A vacancy that has no possibility of holding an electron has little significance as a vacancy.

On the basis of this picture, we can see that the electronegativity must be related both to the effective nuclear charge and to the distance over which that charge must act. We have seen that the effective nuclear charge increases from left to right across a period of major group elements in the periodic table. We have also seen that the atomic radius decreases with increasing effective nuclear charge, because, as the latter increases, it is capable of holding electrons closer together despite their mutual repulsions. If electronegativity is proportional to the coulombic force between the effective nuclear charge and an electron in the outermost principal quantum level, then **both changes,** the increase in effective nuclear charge and the decrease in radius, **contribute to an increase in electronegativity.** Thus we

have a very satisfactory qualitative explanation of the manner in which electronegativity changes across the periodic table.

Lacking an exact method of determining electronegativities of atoms, we
must find values that are consistent with our qualitative views and also that can be
validated by successful quantitative application. The alternative, of denying the
validity of electronegativity simply because we cannot yet measure it precisely,
seems totally unacceptable. First, it seems doubtful that anyone would reject the
concept of polar covalence arising from initial differences among atoms of different
elements. It would seem equally unreasonable for one to deny that the attraction
between the positive charge of the nucleus and the negative charge of an electron in
the outermost principal quantum level is the only possible cause of the tendency
for the electron to remain there as part of the cloud instead of being expelled by
the remainder of the cloud. All we are doing here is calling that attraction
electronegativity. We are also saying that bond polarity is caused when one of the
bonding atoms has a greater electronegativity than the other. **The initially more
electronegative atom will become partially negative at the expense of the other
atom, and the extent of unevenness of sharing will be related to the magnitude of
the electronegativity difference.**

Electronegativities from the Compactness of Atoms. The evaluation of
electronegativity, which originally led to the further developments of equalization,
partial charge, and the quantitative description of polar covalence, began as an
effort to understand the striking differences between the atoms of the M8 family
and ions isoelectronic with those atoms. Students of elementary chemistry have
long been indoctrinated with the idea that a magical stability is associated with the
number of electrons of an M8 element such that other atoms tend to acquire that
number by gain or loss of electrons. The implication that the other atoms achieve
the same magical stability by acquiring the magical electron number is widely
assumed but of course is completely inaccurate. In any series isoelectronic with an
M8 element, positive ions and negative ions differ in opposite directions from the
M8 neutral atom in a very significant manner. Positive ions tend to attract and
negative ions tend to lose or share their electrons in a variety of chemical phenomena, whereas the neutral atoms do neither. In other words, there is magic in the
number of electrons of an M8 element *only* because that number is **equalled by the
nuclear charge.**

During my first year of teaching, I was explaining to my students of general
chemistry that iodine, with 53 electrons, tends to acquire one electron so that it
can achieve the number 54 and in this sense resemble xenon. In one of those rare,
lucid moments that come once or twice in a lifetime even to the least of us, I
suddenly realized that the resemblance between iodide ion and xenon atom in fact
ends where it begins—with 54 electrons. In striking contrast to the xenon atom, the
iodide ion is a very reactive chemical species. In search for the fundamental
difference, other than nuclear charge, between the two species, I investigated the

fact that iodide ion would be expected to be substantially larger than xenon atom since its electronic cloud is restrained by a nuclear charge of only 53 protons instead of 54. This led to the recognition that the relative compactness of an atom, as measured in terms of average number of electrons per unit volume of the atom, seems to parallel the electronegativity. Atoms having relatively dense electronic spheres are of relatively high electronegativity, tending chemically to acquire electrons to cause the density to decrease. Atoms having relatively attenuated electronic spheres are of relatively low electronegativity, tending chemically to lose electrons causing the density of the remaining cloud to increase.

Qualitatively at least, the fundamental concept seemed sound. Atoms that do not have very compact electronic spheres thereby reveal their inability to hold their own electrons tightly together, and there is no reason to expect them to exert much attraction for any electron from somewhere else. On the other hand, any atom that can hold its own electrons closely together despite their mutual repulsion must, if it has an outermost vacancy, be able to attract an electron from somewhere else into that vacancy with significant power.

To the extent of availability of appropriate data, therefore, there was calculated for each element an "average electronic density," expressed mathematically as

$$D = Z/4.19r^3$$

where Z is the atomic number (number of electrons in the cloud) and 4.19 is $\frac{4}{3}\pi$, and r is the nonpolar covalent radius. Although this kind of evaluation gave results approximately paralleling the accepted Pauling electronegativities, it was desirable to know the relative compactness of the M8 elements. Moreover, there were some obvious discrepancies, such as iodine of lesser electronegativity having a higher D value than chlorine.

Pauling had estimated M8 radii by interpolation of the radius versus nuclear charge but using the calculated "univalent ionic radii." It seemed somewhat more appropriate to use the "real" ionic radii, which give two curves, one for anions and the other for cations, and to try to extrapolate where they might meet. This gave a somewhat different set of radii of the M8 atoms. It later became evident that these radii are actually hypothetical radii of "ions having zero charge" and thus not directly comparable to other types of radii. However, they were useful in calculating electronic densities of the M8 atoms, which were found to be low relative to most of the active elements but variable within the group. If relative compactness varies with electronegativity, what was the meaning of variations in elements having no electronegativity?

It seemed necessary to make a correction for the variation of electronic density with atomic number, independent of electronegativity. Therefore the correction was incorporated into an evaluation of electronegativity S:

$$S = D/D_i$$

where D_i is the average electronic density corresponding to a given atomic number, obtained by linear interpolation of the D values calculated for the M8 elements.

This measure of electronegativity, originally named (unfortunately, it is now believed) "stability ratio," gave the values listed in Table 2-1, which are essentially the values still in use. A few have been modified somewhat as shown in the table, which also lists the values currently favored. The basis for modification is given in the discussion of the properties of each individual element (Chapter Three). It is

TABLE 2-1
Electronegativities of the Elements from Relative Atomic Compactness

Element	Original value[a]	Revised value
H	3.55	3.55
Li	0.74	0.74
Be	1.91	1.99
B	2.84	2.93
C	3.79	3.79
N	4.49	4.49
O	5.21	5.21
F	5.75	5.75
Na	0.70	0.70
Mg	1.56	1.56
Al	1.94	2.22
Si	2.62	2.84
P	3.34	3.43
S	4.11	4.12
Cl	4.93	4.93
K	0.56	0.42
Ca	1.22	1.22
Zn	2.84	2.98
Ga	3.23	3.28
Ge	3.59	3.59
As	3.91	3.90
Se	4.25	4.21
Br	4.53	4.53
Rb	0.53	0.36
Sr	1.10	1.06
Cd	2.59	2.59
In	2.86	2.84
Sn	3.10	3.09
Sb	3.37	3.34
Te	3.62	3.59
I	3.84	3.84
Cs	0.49	0.28
Ba	1.02	0.78
Hg	2.93	2.93
Tl	3.02	3.02
Pb	3.06	3.08
Bi	3.14	3.16

[a]From R. T. Sanderson, "Chemical Periodicity," Van Nostrand-Reinhold, Princeton, New Jersey, 1960.

surely more than coincidence that the electronegativities first evaluated and published in 1954 are identical with or only slightly changed from those ten years later found indispensable for the accurate calculation of bond energy. (See also Fig. 2-2.)

The discussion will return to electronegativity presently, but first we must examine another aspect of effective nuclear charge.

Homonuclear Single Covalent Bond Energy

The energy released when two like atoms unite through a single covalent bond in which two electrons, one from each atom, are exactly evenly shared by the two nuclei is the homonuclear single covalent bond energy. For a diatomic molecule of an element, in which the bond is a single bond, it is the same as the bond dissociation energy. For a network solid, held together by single bonds, it is the average energy per bond as calculated from the experimental atomization energy of the solid.

The presumption is that the homonuclear bond energy represents the net attraction involved when two like nuclei are equally attracted to the same pair of electrons. This must be a coulombic attraction and thus involves coulombic energy. The *average* location of electrons shared between two like atoms must be midway along the bond axis between the two nuclei. In other words, the bond energy must be proportional to the coulombic energy of interaction between the nucleus and a

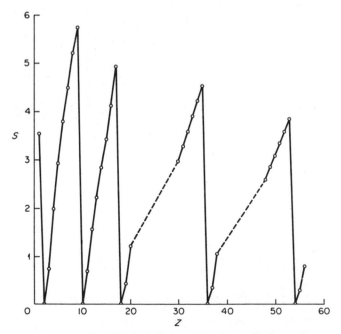

Fig. 2-2 Periodicity of electronegativity.

pair of electrons at the distance of the nonpolar covalent radius. Since the nuclear charge is mostly shielded by the intervening electrons, it is the effective nuclear charge that determines the bond strength, along with the distance over which the charges must interact. We may then reasonably expect that the **homonuclear single covalent bond energy across the major groups of the periodic table must increase steadily from left to right, because of an increasing effective nuclear charge operating over a decreasing distance.** As nearly as experimental evidence can assure us, this is true through the first four groups. Thereafter there is a surprising weakening of the homonuclear single covalent bond energy, the recognition of which has permitted valuable insight into many common chemical phenomena, as will be explained later. Let us for the moment consider only the idealized situation, in which there is no such weakening and the predicted trend persists as expected.

Eliminating the Effective Nuclear Charge

It has been shown that the radius of an atom is strongly influenced by the effective nuclear charge, as are also the electronegativity and the homonuclear single covalent bond energy. Consequently, it is frustrating to be unable to evaluate accurately the effective nuclear charge. However, there is, in a manner of speaking, a way out. We can eliminate the urgency of the need for knowing the effective nuclear charge.

Remember that electronegativity can be thought of as proportional to the coulombic attraction between an electron at the distance of the covalent radius and the effective nuclear charge. More recently it is pointed out that the homonuclear single covalent bond energy can be thought of as proportional to the coulombic energy between an electron at the distance of the covalent radius and the effective nuclear charge. It follows that electronegativity and homonuclear single covalent bond energy are proportional to each other. Coulombic *attraction* is measured as charge product divided by distance squared. Coulombic *energy* is measured as charge product divided by distance. Consequently, the homonuclear single covalent bond energy is proportional to the product of the radius and the electronegativity

$$E \propto \frac{Z_{\mathrm{eff}}e^2}{r} \qquad S \propto \frac{Z_{\mathrm{eff}}e^2}{r^2} \qquad E = CrS$$

where C is a proportionality constant.

To test this simple relationship, the rS product was plotted against the experimental value of E for all the major group elements for which data could be obtained. The series of straight lines shown in Fig. 2-3 resulted, the lines all converging at the origin. It appeared that a definite relationship among the slopes should exist. Indeed it does, such that the proportionality constant C can be expressed as a function of n, which in turn denotes the electronic type of structure. These results are summarized in Table 2-2.

It must be pointed out that even if they are derived from meaningful values of n, the proportionality constants C are only empirically determined. If the values

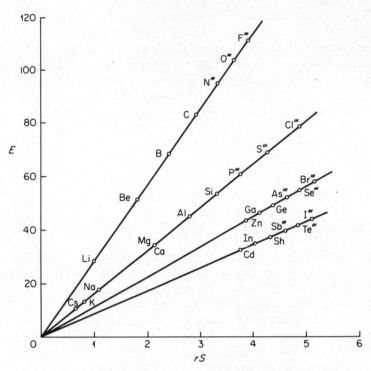

Fig. 2-3 Relationship of homonuclear bond energy to rS product.

TABLE 2-2

**Relationship between Electronegativity,
Homonuclear Bond Energy, Nonpolar Covalent
Radius, and Electronic Configuration $E = CrS^{a,b}$**

n	Electronic type of atom	C
2	Two electrons in penultimate shell	28.5
3	Eight electrons in penultimate shell	16.1
4	Period 4: 18 electrons in penultimate shell	11.2
5	Period 5: 18 electrons in penultimate shell	8.6

[a] R. T. Sanderson, *J. Inorg. Nucl. Chem.* **28,** 1553 (1966).

[b] Equations:

$$C = \frac{37.0}{n - 0.70}; \quad E = \frac{37.0rS}{n - 0.70}; \quad S = \frac{E(n - 0.70)}{37.0r}$$

of C are assumed correct, or at the least, self-consistent, then of course the relationship $E = CrS$ allows a semiexperimental determination of the electronegativity, since both the radius and the bond energy are experimental quantities. Where these quantities are accurately known, the relationship permits correction of electronegativity values, as detailed later.

"Lone Pair Bond Weakening" Effects

The elements nitrogen, oxygen, and fluorine have long been known to exist in the elemental state as diatomic gas molecules N_2, O_2, and F_2. The bond energies in these have been determined as 226, 119.2, and 37.8 kpm, respectively. But these are recognized to correspond to a triple bond for nitrogen, a double bond for oxygen, and a single bond for fluorine. Evidence for anomaly exists for all three, but it is most obvious for fluorine with the single covalent bond. The measurement of the dissociation energy of F_2 is experimentally difficult and still the basis for some controversy. A series of independent investigations some years ago seemed to establish the dissociation energy of F_2 as below 40 kpm, whereas the values for Cl_2, Br_2, and I_2 were well known to be about 58.2, 46.1, and 36.1 kpm. Other properties of the halogens appear much more consistent in alignment from fluorine through iodine, and the low dissociation energy of F_2 was very much unexpected. Extrapolation procedures all would lead to values much higher for fluorine than for chlorine.

The single covalent bond energy for nitrogen and oxygen could not so readily be obtained except through indirection. However, a fortuitously nearly correct assumption is that the H—O bonds in water and in hydrogen peroxide are essentially equal in energy. Similarly, the H—N bonds in ammonia and hydrazine are of equal energy. Therefore it is possible to subtract the H—O energies from the experimental atomization energy of hydrogen peroxide to learn by difference the O—O energy, which is about 34 kpm. Similarly, the N—N energy in hydrazine is found to be about 39 kpm. A discrepancy thus exists in the bond energy attainable for nitrogen, oxygen, and fluorine. Chemists were quite familiar with multiple bonds in hydrocarbons and had learned that double bonds are never twice as strong as single bonds or triple bonds three times as strong as single bonds. For example, the energy of a C—C bond is about 83 kpm, half that of a C=C bond is about 73 kpm, and one-third of a C≡C bond is about 64 kpm. Consequently it seems quite extraordinary that half the double bond energy of O=O is about 60 kpm, whereas the single bond energy is only about 34. And how could one-third of the triple bond energy in N_2 be 75 kpm when the single bond energy is only 39?

Various explanations were offered, principally by Pitzer[6] and by Mulliken,[7] based essentially on the concept that, as demonstrated in carbon chemistry, when two atoms of nitrogen, oxygen, or fluorine come close enough to form a single

[6] K. S. Pitzer, *J. Amer. Chem. Soc.* **70**, 2140 (1947).
[7] R. S. Mulliken, *J. Amer. Chem. Soc.* **72**, 4493 (1950); *ibid.*, **77**, 884 (1955).

covalent bond, other orbitals are also close enough to interact. In carbon the π orbitals can form double or triple bonds, but in nitrogen and oxygen and fluorine there are too many electrons, so these lone pair electrons repel one another, weakening the single bond and causing it to have a lower than expected bond energy. In other words, the phenomenon was recognized but ascribed to an interatomic effect, that of repulsions between nonbonding electrons on adjacent atoms.

As a result of the $E = CrS$ relationship, the picture has changed. As mentioned earlier, when the energy versus radius electronegativity product is linearly extrapolated, nothing like the low single bond energies ascribed to nitrogen, oxygen, and fluorine are obtained. Instead, the values steadily increase: lithium 28, beryllium 51, boron 69, carbon 83, nitrogen 95, oxygen 104, fluorine 111. These extrapolated values were assumed to represent what the single covalent bond energy for these elements should be.

To test these values, it was first necessary to learn the relationship between single and multiple covalent bond energies. A study of double bonds and of triple bonds in a series of hydrocarbons revealed that a double bond is 1.50 times as strong as a single bond would be if it were of the same length as the double bond. Similarly, a triple bond is about 1.75 times as strong. The empirical multiplicity factor for aromatic rings is $1 + 0.33n$, where n is the number of π electrons per bond. We may generalize with reasonable accuracy for all multiplicity by the empirical equation $m = 0.74p + 0.57$, where m is the multiplicity factor and p is $n^{1/3}$. These factors can be used to calculate from a multiple bond energy what the normal single bond energy should be. Conversely, if we think we know what the single bond energy should be, we can use it in calculating a multiple bond energy.

For nitrogen, the extrapolated homonuclear single bond energy is 94.8 kpm. The bond length in N_2, which has a triple bond, is 1.10 Å, compared to a nonpolar covalent radius sum of 1.48 Å. The bond energy E is then equal to the single bond energy times the factor 1.75 and corrected for the shorter distance:

$$E = 94.8 \times 1.75 \times 1.48/1.10 = 223.5 \text{ kpm}$$

This is in excellent agreement with the experimental dissociation energy of N_2, which is 226.0 kpm. One may conclude that the bond weakening evidenced in a single bond energy of 39 kpm does not occur when a triple bond is formed.

An example of a triple bond involving oxygen is in carbon monoxide. Although this bond is not homonuclear, it is assumed that the triple bond factor 1.77 applies to both covalent and ionic energy contributions. By using the unweakened single bond energy extrapolated for oxygen, 103.9 kpm in the calculation, we find the calculated dissociation energy to be about 260 kpm, agreeing very well with the experimental value of 257 kpm.

What of the double bond energy in O_2? Experimentally, this is known to be 119.2 kpm. Since this is lower than the double bond energy for C=C and we should expect a higher value, we assume that this bond must be weakened to some extent.

We therefore calculate the single bond energy that corresponds to the double bond energy of oxygen, corrected for length:

$$E = \frac{119.2 \times 1.21}{1.50 \times 1.40} = 68.7 \text{ kpm}$$

The difference between the "unweakened" energy, 103.9, and 68.7 is 35.2. If the weakening effect is assumed to be completely removed in a triple bond and completely operative in a single bond, we may assume that in a double bond the weakening effect is halfway between. To test this assumption, we subtract 35.2 from 68.7 and get 33.5, nearly the same as the single bond energy determined from the thermochemistry of water and hydrogen peroxide.

Thus are found three different single covalent bond energies for oxygen. The unweakened energy is 103.9, designated E'''. The partly weakened energy E'' is 68.7, and the fully weakened energy E' is 33.5 kpm.

Speculation as to Cause. Before continuing to develop these ideas and their applications, we need to inquire what is the cause of the bond weakening. The most obvious possibility is not necessarily the real one, but we have here at least an obvious electronic difference that might account for our observations. Easily recognized is a feature of elements beyond Group M4 that is not present in the first four groups. In M5 and thereafter, the outer shell contains lone pair electrons in addition to those used in normal covalence. The manner in which such a lone pair might weaken a covalent bond is not yet understood, and it is therefore not certain whether the lone pair electrons are in fact involved. However, the effect seems consistent with such a concept, in that the weakening effect appears to become diminished under conditions in which lone pair interference with the bonding might reasonably be expected to be reduced.

For example, the triple bond in N_2 can be pictured as concentrating six bonding electrons within the internuclear region, so high a concentration being capable of excluding the lone pair electrons and in fact repelling them to the far sides of the molecule where they are no longer able to exert any weakening effect. The formation of a double bond, as in oxygen, can similarly be imagined as decreasing the weakening effect of a lone pair but without destroying it completely. In this way we can rationalize the involvement of the E''' energy in N_2, the E'' energy in O_2, and the E' energy in F_2.

The earlier explanations of this weakening effect, as the result of interelectronic repulsions between lone pair electrons on separate atoms, become void when we consider such compounds of these elements as are formed, for example, by hydrogen. The bond energies in ammonia, water, and hydrogen fluoride can be calculated very accurately, as detailed in Chapter Six. In NH_3, H_2O, and HF the E' energy is applicable. This proves that the weakening effect cannot be an interatomic effect, as earlier explained, because there are certainly no lone pair electrons on the hydrogen atom to repel the lone pair electrons of the nitrogen, oxygen, or fluorine.

Instead, the correct description of the weakening effect is that it is an *intra*atomic effect, caused and manifested within the single atom.

Reduction of Weakening Effect. From the specific examples discussed above, we may generalize that the E''' energy is involved wherever a triple bond involving an M5, M6, or M7 element is formed and that the E'' energy is applicable to all such double bonds. This generalization is verified by an abundance of experimental and calculated evidence.

But what can be the significance of stronger than "normal" single homonuclear bond energies in fluorine, which cannot form more than one single bond per atom? Here comes a second revelation concerning the so-called lone pair bond weakening effect. There are, in fact, many examples of single covalent bonds in which part or all of the weakening effect is removed. In other words, single covalent bonds are common in which the single homonuclear bond energy of one of the two elements involved is evidently either only half-weakened or not weakened at all. For example, the atomization energy of $AlF_3(g)$, determined from experiment as 421 kpm, is significantly higher than would be calculated using the fluorine homonuclear energy of 39.5 kpm. Application of the unweakened energy for fluorine, 111.4 kpm, however, gives an atomization energy of 428 kpm, in reasonably good agreement with the experimental value.

This phenomenon is in need of much further study. At the present time, it appears that reduction of bond weakening occurs under the following conditions: (1) In a polar heteronuclear single covalent bond it is observed only in the homonuclear covalent energy of the more electronegative atom, even when both atoms are of groups M5, M6, or M7. (2) It occurs only when the originally less electronegative atom has outermost vacant orbitals that might be able to accommodate extra electrons. (3) An exception to (2) occurs when the atom is already involved in a double bond to another atom, as exemplified by the single C—O bond in a carboxyl group or the single C—N bond in an amide group, in which the E'' energy of oxygen or nitrogen is applicable instead of the E' energy. (For example, the bridging oxygen atom in an ester is attached to the carboxyl carbon by a single bond using the O'' energy and to the carbon of the former alcohol by a single bond that is weaker, using the O' energy).

A tentative and partial explanation is that the lone pair weakening effect arises from lone pair electrons intervening between bonding electrons and nucleus. Otherwise vacant orbitals on the other atom may somehow accommodate the lone pair electrons in such a manner as to reduce their weakening effect without adding to the multiplicity of the sharing or in any other way strengthening the bond. For example, in SiF_4 the withdrawal of substantial charge by the fluorine atoms from the silicon atom would heighten the interaction between the silicon nucleus and the silicon outermost d orbitals, such that the latter would become available for attracting lone pair electrons from the fluorine atoms in such a way as to eliminate their weakening effect. Thus the full fluorine bond strength is realized in the compound, $Si—F'''$. Many other examples will be discussed.

The reason for believing that the lone pair weakening effects must be reducible without increasing the multiplicity of the bond is that this reduction of weakening affects only the covalent contribution to the total bond energy. The total bond energy consists of a blend of covalence and ionicity, as will be detailed later, and the multiple bond factors for double and triple bonds apply to both the covalent and the ionic contributions.

The Heavier Elements of Groups M5, M6, and M7.   Extrapolation of the $E = CrS$ relationship to provide larger homonuclear single bond energies is not limited to finding values for nitrogen, oxygen, and fluorine. These appear of most immediate interest because the weakening effects in these elements have long been recognized. What has not been generally recognized is that the heavier elements of the same groups are also susceptible to this phenomenon, although much less so. For example, chlorine has a well-known dissociation energy of 58.2 kpm, but its unweakened homonuclear single covalent bond energy is 78.6 kpm. The anomalous series of E' energies of the halogens, F 39, Cl 58, Br 46, and I 36 kpm attracts our attention at once, but the situation appears quite different when we compare their unweakened energies E''': F 111, Cl 79, Br 58, and I 44 kpm.

The thermochemical data available for some of the heavier elements of these groups are somewhat limited but appear adequate to establish at least approximately the values of Table 2-3, which lists homonuclear single covalent bond energies for all the elements of M5, M6, and M7.

Return to Electronegativity

The above discussion should clarify and emphasize what has not been sufficiently appreciated in the past—that the concept of electronegativity is far more

TABLE 2-3

**Weakened and Unweakened
Homonuclear Single Bond Energies**

Element	E'	E''	E'''
N	39.2	67.0	94.8
O	33.5	68.7	103.9
F	39.5	75.5	111.4
P	51.1	55.9	60.7
S	55.0	62.1	69.0
Cl	58.2	68.4	78.6
As	43.4	47.7	52.0
Se	44.0	49.3	54.6
Br	46.1	52.0	58.0
Sb	32.0	35.8	39.6
Te	34.	38.	41.7
I	36.1	40.0	43.9

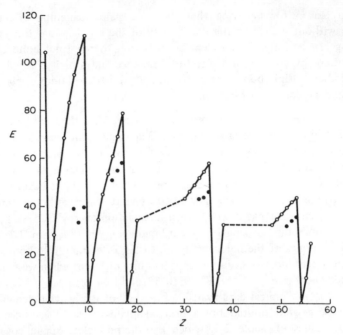

Fig. 2-4 Periodicity of homonuclear bond energy. (Points below curve represent weakened energies.)

broadly based and potentially useful than merely to tell us qualitatively the direction and extent of bond polarity. In particular, we see that **it is as powerful a factor in homonuclear bonding as in heteronuclear bonding.** Perhaps others would alter the emphasis, but, in my opinion, the big "one-two" of atomic properties are (1) **outermost configuration** and (2) **electronegativity.**

While we await the completion of the marble palace, a wooden shelter will protect us from the weather and bring us comfort. Whether or not he will admit it, since the dawn of history man has been moving from one makeshift to another, hoping one day for the ultimate in housing. In this area we encounter three schools of thought, and it may be useful here to recognize these schools and identify our own preference. Those of one school appear satisfied that we have already reached the state of "good enough" and prefer to ignore its imperfections as though they were not obvious. Those of another school scorn anything less than perfection. They prefer shivering in the clammy rain that pours into the beginnings of the roofless marble palace to compromising their scientific ideals for the sake of practical comfort. Between those two schools, each admirable in a limited way, lies the school of thought preferred by this author and, I hope, most readers. We gaze wistfully at the vision of ultimate perfection in the structure of our knowledge and, fascinated by the challenge of our dream, work long and hard toward achieving even some slight measure of it. Nevertheless, we are realists and practical and are willing to accept, at least tentatively, less than perfection while we await the real thing.

In a sense, electronegativity is a relatively crude makeshift, an oversimplifying concept that operates best in the realm of the practical. However, the following aspects of electronegativity should be appreciated:

(1) It is not imaginary but represents a very real and significant quality of an atom.

(2) The values assigned to it may not be precisely perfect, but they can be extraordinarily useful. They play a vital role in the quantitative description of polar covalence, allowing the correct prediction not only of bond energy but also, in many instances, of bond length. Electronegativity has unique and valuable applications to the interpretation of a vast area of common chemistry.

Although enthusiasm is a very natural and human response to a pleasing approach toward understanding, we need to remain keenly aware that what we really seek is to understand bonds as they are, not necessarily as we might like them to be. Be assured that, in the discussion to follow, no facts or arguments will be consciously withheld just because they might seem unfriendly.

THREE

Selected Values
of Atomic Properties

INTRODUCTION

The special importance of the nonpolar covalent radius, homonuclear single covalent bond energy, and electronegativity of an element have already been discussed. The empirical relationship $E = CrS$, where C is a proportionality constant related to the electronic configuration of the atom, has also been considered. By judicious application of this relationship and broad consideration of the interrelationships among the chemical elements as revealed by their physical and chemical properties, it has been possible to select "best" values of these atomic properties, which have then been applied consistently in the calculation of bond energies. Only a few minor changes have been made since the first edition of this book. Since there is no absolute basis, as yet, for establishing the validity of some of these properties for some of the elements, the complete details of selection are presented here so that the derivation or origin of each value becomes clear.

MAJOR GROUP ELEMENTS

Hydrogen

The experimental dissociation energy of H_2 at 25° is well established as 104.2 kpm.[1] The electronegativity value, 3.55, was originally somewhat arbitrarily as-

[1] Unless otherwise specified, the thermochemical data used in this book are taken from one of the following sources: "Selected Values of Chemical Thermodynamic Properties,"

signed as believed to be of the correct order of magnitude, intermediate between boron and carbon but nearer to carbon than to boron. No basis for seriously questioning this value has arisen. On the contrary, it seems to provide very reasonable values of partial charges in hydrogen compounds, which in turn are applicable to successful bond energy calculations.

Normally an acceptable nonpolar covalent radius can be evaluated as half the length of a nonpolar bond between uncharged like atoms, but not for hydrogen. The bond length in the H_2 molecule, 0.74 Å, corresponds to a nonpolar covalent radius of 0.37 Å. In fact, however, this value is unsatisfactory to describe bond lengths in hydrogen compounds, wherein 0.32 Å appears more acceptable. If the homonuclear bond energy were corrected for this shorter length, it would be 120.6 instead of 104.2 kpm. This higher value does not seem applicable to bond energy calculations. There are thus numerous aspects of hydrogen properties that remain puzzling and are treated somewhat arbitrarily. Nevertheless, hydrogen compounds are dealt with quite accurately based on the following hydrogen properties: homonuclear single covalent bond energy, 104.2 kpm; nonpolar covalent radius, 0.32 Å; and electronegativity, 3.55.

Lithium

Near the boiling point of lithium, a small fraction of its mainly monatomic vapor consists of dimers, Li_2 molecules, whose properties have been measured. The dissociation energy is 26.5 kpm. The nonpolar covalent radius of 1.34 Å is half the experimental bond length. For $n = 2$, the constant C in the equation $E = CrS$ is 28.5. To fit this relationship exactly, the electronegativity would have to be 0.69. The original assigned value was 0.74, which is more consistent with the expectedly lower sodium value of 0.70. This would correspond to a bond energy of 28.3 kpm. There being no basis for choice, both values appearing quite satisfactory in the treatment of lithium compounds, a compromise is adopted, with lithium assigned the experimental dissociation energy of 26.5 kpm but the original electronegativity of 0.74, with the nonpolar covalent radius 1.34 Å.

Circular No. 500, U. S. Bureau of Standards (1949); *Nat. Bur. Stand. (U. S.) Tech. Note* No. 270–3 (1968) and No. 270–4 (1969); Joint Army, Navy, and Air Force Thermochemical Tables (JANAF), Dow Chemical Co., Midland, Michigan; T. L. Cottrell, "The Strengths of Chemical Bonds," Butterworths, London, 1958.

Bond lengths and other structural data were taken largely from the following: A. F. Wells, "Structural Inorganic Chemistry," 3rd Ed., Clarendon Press, Oxford, 1962; "Interatomic Distances," Special Publication No. 11, The Chemical Society, London, 1958; "Interatomic Distances Supplement," Special Publication No. 18, The Chemical Society, London, 1965.

Beryllium

Be_2 molecules, if they exist, appear to be extremely unstable and unsuitable for determining homonuclear single covalent bond energy. The value adopted in the first edition has appeared on further study to be unreasonably high, so here a significant change has been made. The homonuclear single covalent bond energy of 51.6 kpm has been adopted, along with the revised electronegativity of 1.99 (instead of the former 2.39) corresponding to the $E = CrS$ relationship. The nonpolar covalent radius is taken as 0.91 Å.

Boron

From the standard heats of formation of $B(g)$ and $B_2(g)$ can be calculated a B—B energy of about 70.5 kpm. There is no assurance that this bond represents exactly the more typical bonding condition in which boron forms three covalent bonds, but it appears similar. The bond length is a little less than twice the accepted nonpolar covalent radius 0.82 Å, being 1.59 Å. Correcting the experimental bond energy to the more normal bond length gives 68.4 kpm. To fit the $E = CrS$ equation, this requires a slight upward revision of electronegativity from an original 2.84 to 2.93. Satisfactory values for boron are then 68.4 kpm for the homonuclear single covalent bond energy, 0.82 Å for the nonpolar covalent radius, and 2.93 for the electronegativity.

Carbon

Small but doubtless significant difficulties are recognized in the determination of certain carbon properties. There is no reason to suspect carbon atoms of being unique in this respect, and it is probable that the same difficulties occur for other kinds of atoms too but have been successfully ignored. As will be discussed in more detail in Chapter Eleven, our attention is more likely to be drawn to minor factors when studying the chemistry of carbon atoms, because their effects are multiplied by the number of carbon atoms in the molecule. The fact that these effects begin to become apparent in carbon chemistry suggests that the methods of bond energy calculation described herein are possibly very nearly as accurate as they are capable of being from a practical viewpoint. It may be predicted that further refinements will for the most part encounter much greater complexities whose disadvantages may outweigh any advantages of being a little more precise or less approximate.

The atomization energy of graphite has not been easily accessible through experiment, and some disagreements have arisen. However, it is now generally accepted as about 171.3 kpm. The mechanical process of atomizing graphite is not easily visualized, but it is known that the standard heat of formation of diamond from graphite is 0.45 kpm and the diamond structure is more easily treated. The atomization energy of diamond is therefore 170.8 kpm. Atomization of the diamond structure requires the equivalent of the breaking of two single covalent bonds per carbon atom, so one single covalent bond energy should be 85.4 kpm. Thermochemical data for hydrocarbons having more open structures of carbon atoms than diamond appear to favor a somewhat lower C—C single bond energy, around 83 kpm. By back calculation from the experimental bond energies of some of its simpler compounds, we find the carbon bond energy to be as follows: CO_2, 83.2; CO, 80.6; CS_2, 81.3; CH_4, 84.0 kpm.

The electronegativity of carbon as determined by the relative compactness method is 3.79, a value that seems very reasonable and consistent with other properties of carbon. The bond length in diamond as well as in many organic compounds is 1.54 Å, leading to a nonpolar covalent radius of 0.77 Å. In the $E = CrS$ equation, these values of r and S correspond to 83.2 kpm, which is adopted for the C—C energy.

Nitrogen

The electronegativity of nitrogen by the relative compactness is 4.49. The nonpolar covalent radius is 0.74 Å. By assuming the experimental heat of formation of hydrazine to be correct, we can back-calculate to find the N—N bond energy, assuming the N—H bonds to be the same as in ammonia. This gives us the adopted N' energy of 39.2 kpm. The completely unweakened bond energy, as extrapolated from the $E = CrS$ equation, is 94.8, which is taken as the N''' single bond energy. Halfway between the two, 67.0 kpm, is the N'' value.

Oxygen

Oxygen and fluorine appeared anomalous as reported in the first edition and have continued to cause problems. These have now been minimized, if not removed, by assuming that the fully weakened single bond length might well be somewhat longer than twice the nonpolar covalent radius and adjusting the radii downward, for oxygen to 0.70 and for fluorine as noted below. This called for minor changes in the homonuclear bond energies, the $E = CrS$ relationship, combined with the compactness electronegativity of 5.21 and the radius of 0.70 Å, giving 103.9 for O'''. We then get 68.7 kpm from the O_2 molecule data and evaluate the O' energy as 33.5 kpm. Reduction of radius and increase in unweakened bond energy tend to cancel one another in effects on calculated bond energies, so results are not very different from those obtained earlier. However, the bond lengths appear more satisfactorily represented.

Fluorine

As described for oxygen, a minor revision of nonpolar covalent radius from 0.71 to 0.68 Å leads to better bond length calculations and to slightly modified bond energies. As for oxygen, the triple prime energy is now much closer to the value extrapolated from the $E = CrS$ equation. The electronegativity is unchanged, at 5.75, but the homonuclear single covalent bond energies now become F', 39.5; F'', 75.5; and F''', 111.4 kpm. Some of the triple prime energy calculations are made poorer, but, on the average, the results appear as satisfactory as those obtained prior to this modification.

Sodium

The vapor of sodium resembles that of lithium in containing a small concentration of Na_2 molecules near the boiling temperature. The dissociation energy is measured as 18.0 kpm, taken as the homonuclear single covalent bond energy. Half the bond length is 1.54 Å, the nonpolar covalent radius. The $E = CrS$ equation would then bring the electronegativity to 0.72, hardly worth changing from the original value of 0.70, which is retained.

Magnesium

The nonpolar covalent radius of magnesium is 1.38 Å and the electronegativity by the relative compactness method is 1.56; where $n = 3, E = 16.1, rS = 34.6$ kpm.

Since no Mg—Mg bond energies are known from experiment, the above bond energy of 34.6 must be tested by calculation of the atomization energies of magnesium compounds. Here a difficulty is encountered, in that nearly all calculated values tend to be too high by an average of about 10 kpm. Although it is possible to adjust both E and S together, several attempts to find values giving better agreement with experimental atomization energies failed. This suggested the possibility of a consistent error in the determination of the experimental atomization energies, which are the difference between the sum of the atomization energies of the separate elements and the standard heat of formation of the compound. The only value in common is the atomization energy of magnesium. This is presumably reliably reported as about 35.5 kpm.

There is really little other justification than the bond energy calculations for suspecting this value, except that it is out of line in M2, being lower than for calcium. However, *if* the atomization energy of magnesium were really about 45.5 kpm instead of 35.5, then the following calculated and experimental atomization energies result: MgF(g), 130.6, 132.0; MgF_2(g), 262.2, 258.9; $MgCl_2$(g), 203.2,

204.3; $MgBr(g)$, 84.6, 85; $MgBr_2(g)$, 177.2, 173.4; $MgF_2(c)$, 358.4, 352.0; $MgCl_2(c)$, 256.7, 256.9; $MgBr_2(c)$, 224.2, 222.9; $MgI_2(c)$, 180.0, 180.0; $MgO(c)$, 245.8, 248.9 kpm.

Tentatively, magnesium is accepted as conforming with the $E = CrS$ relationship if its radius is 1.38, its electronegativity 1.56, and its bond energy 34.6 kpm, provided the atomization energy of magnesium metal is indeed some 10 kpm higher than literature values indicate.

Aluminum

Taking the electronegativity of aluminum as 2.22 and the nonpolar covalent radius as 1.26 Å, then from $E = 16.1\ rS$ for $n = 3$ is obtained the homonuclear bond energy parameter of 45.0 kpm.

Since no experimental bond energy for Al—Al is available, some indirect procedure for testing the above value must be used. The aluminum trihalide gas molecules are all assumed to involve the X''' energy for the halogens. On this basis, one can back-calculate the following homonuclear bond energy for aluminum: AlF_3, 44.1; $AlCl_3$, 45.8; $AlBr_3$, 49.9; and AlI_3, 42.6; average, 45.6 kpm. Although some other appropriate combination of bond energy and electronegativity might give acceptable results in bond energy calculations, the above values are tentatively accepted as being very satisfactory for the limited data available for aluminum compounds: $E = 45.0, r = 1.26, S = 2.22$.

Silicon

The bond length in solid silicon, of diamond structure, is 2.34 Å. Half of this value, 1.17 Å, is the nonpolar covalent radius. The electronegativity of silicon is revised slightly upward from the older relative compactness value to 2.84.

The atomization energy of solid silicon is reported to have the experimental value of 108.9 kpm. If as with diamond, half of this is taken as the single bond energy, the homonuclear single bond energy is 54.5. If, however, as also with diamond, extra bond energy comes from having a close assemblage of like atoms, a slightly lower energy might be more appropriate. From the above radius and electronegativity values and the relationship $E = 16.1\ rS$ for $n = 3$, the value of E is 53.4 kpm—a reasonable reduction from the 54.5 comparable to that observed in carbon. The 53.4 value gives very satisfactory bond energy calculations for silicon compounds and thus appears to be acceptable and is adopted here.

Phosphorus

The covalent radius of phosphorus is 1.10 Å, half the bond length in black phosphorus, which is nearly the same as in white phosphorus, P_4. The electronegativity is revised slightly upward from the original relative compactness value of 3.34 to 3.43. The relationship, $E = 16.1\ rS$, gives 60.7 for the phosphorus bond energy.

This, however, must be the value for the P''' energy, since the lone pair on phosphorus must weaken the bond as in nitrogen but, as pointed out earlier, to a much smaller degree.

To check this, let us consider the gaseous molecule, P_2. The bond length is 1.89 Å, instead of the covalent radius sum of 2.20 Å, and its dissociation energy at 25° has been reported as 125.1 kpm. Just as for N_2, one can calculate what single bond energy would apply if there were no lone pair interference or weakening of the bond. The bond is assumed to be a triple bond as in N_2. The corresponding single bond energy is then $(125.1 \times 1.89)/(1.77 \times 2.20) = 60.7$ kpm, in agreement with the calculated value given above. This value is therefore adopted for P'''.

The P' energy can be estimated from data for $P_4(g)$ by subtracting the standard heat of formation of the gas, + 14.1 kpm, from the total atomization energy of 4 moles of red phosphorus (P): $4 \times 79.4 = 318.6$, and dividing by 6, the number of bonds in the P_4 tetrahedron. This gives 50.8 kpm for the P—P bond at the observed bond length of 2.21 Å. Corrected to 2.20 Å this becomes 51.1 kpm. This is the P' bond energy.

Following the established procedure of assuming the P" energy to be midway between the P' and the P''' values, we find for P" the value 55.9 kpm. These values permit bond energy calculations for phosphorus compounds in reasonably good agreement with those determined experimentally. In most applications merely the P' value is needed.

Sulfur

The nonpolar covalent radius of sulfur is taken as 1.04 Å, from a variety of S—S bond lengths ranging between 2.04 and 2.10 Å, and by interpolation of the Z/r^3 function between silicon and chlorine. The relative compactness electronegativity is 4.12. From these two values and the expression $E = 16.1\ rS$, the bond energy is found to be 69.0 kpm. This would correspond to the S''' energy.

Initial attempts to assign a homonuclear covalent bond energy to sulfur based on the experimental atomization energy of $S_8(g)$ gave unsatisfactory results, due to apparently inconsistent thermochemical data in the literature. It was therefore decided to assume that the standard heat of formation of $H_2S(g)$ is completely accurate as reported from experiment. Back-calculation from the experimental atomization energy should give a correct S' energy. The value thus obtained is 55.0 kpm.

The experimentally determined dissociation energy of $S_2(g)$ at 25° is 102.5 kpm. The bond length is 1.89 Å, in comparison to the covalent radius sum of 2.08 Å. The S_2 molecule is paramagnetic like O_2, giving evidence of two unpaired electrons. It therefore seems reasonable to assume that a double bond is present. Using the olefinic double bond factor of 1.50 as in a similar calculation for O_2, one can calculate the S" energy as $(102.5 \times 1.89)/(1.50 \times 2.08) = 62.1$ kpm. Following the usual procedure of considering the S" energy to be halfway between the S' and the S''' energies, we find the S''' energy to be practically the same as calculated above, 69.2 kpm.

From the experimental data, the bond energy in S_8 is calculated to be 63.5 kpm of bonds. This agrees reasonably well with the S'' energy of 62.1. It explains why a suitable S' energy was not obtainable from the thermochemistry of S_8. It also suggests that there is a special reason for the stability of S_8 rings, as will be discussed in Chapter Four.

Chlorine

The bond length in the Cl_2 molecule is known to be 1.99 Å from which the nonpolar covalent radius of chlorine is 0.99 Å. The relative compactness electronegativity is 4.93, a value about which no serious questions have arisen through many applications.

The experimental dissociation energy of Cl_2 at 25° is 58.2 kpm. This value is taken as the Cl' homonuclear single covalent bond energy. In a number of its gaseous compounds, however, chlorine appears to contribute more to the total atomization energy than is indicated by the Cl' energy. Each of these compounds represents a combination of chlorine atoms with another atom that can provide vacant outer orbitals for somehow reducing the bond weakening effect of the chlorine lone pairs of electrons. Back-calculation of the Cl—Cl bond energy from atomization energies of these gaseous chlorides, and assuming the bond energy calculation methods to be correct, gives the following results: $SiCl_4$, 80.8; $GeCl_4$, 80.3; $SnCl_4$, 77.9; BCl_3, 80.4; $AlCl_3$, 75.2; and $BeCl_2$, 72.6. The average is 77.9 kpm, the "experimental" Cl''' energy.

Application of the $E = CrS$ equation where C is 16.1 gives the value of 78.6 kpm for the Cl''' energy. This agrees very well with the experimental value and is adopted. The Cl'' energy halfway between is then 68.4 kpm.

Potassium

Like the other alkali metals, potassium in the vapor state is slightly associated to K_2 molecules. The bond length is 3.92 Å, from which the nonpolar covalent radius is 1.96 Å. The electronegativity is adjusted slightly downward from an earlier value to 0.42. For $n = 3$, which has been found to apply to all elements having a penultimate shell of eight electrons whether in period 3 or not, C is 16.1. The $E = CrS$ equation then gives 13.2 as the bond energy. This is also the experimental dissociation energy of $K_2(g)$.

Calcium

The nonpolar covalent radius of calcium is somewhat uncertain but is taken as 1.74 Å, which appears satisfactory. The electronegativity is 1.22.

No direct experimental evaluation of a Ca—Ca bond energy is known. Some indirect means must be sought. Only gas phase data for the fluorides are known for calcium, strontium, and barium. In these compounds the covalent energy is too small to serve reliably as a basis for back-calculating the metal—metal energy. Back-calculations from the experimental atomization energies of the solid halides are complicated by the fact that different crystalline structures occur. Only CaF_2 has the fluorite structure. Both the chloride and the bromide crystallize in a deformed rutile structure in which each metal atom has two nearest neighbors and four more appreciably farther distant. The iodide has the cadmium iodide layer structure. The bond energy parameters for calcium calculated from CaF_2 (36.8) and CaI_2 (35.7) average 36.3 kpm. Complications in the chloride and bromide calculations suggest that they be omitted from the present evaluation for later discussion.

Application of the $E = CrS$ equation gives a homonuclear single covalent bond energy of 34.2 kpm. Since this appears to be approximately correct, it is tentatively adopted.

Zinc

The nonpolar covalent radius of zinc extrapolated from bromine and germanium using the Z/r^3 function is 1.30 Å. Available thermochemical data for zinc compounds do not provide a very satisfactory basis for evaluating the homonuclear single covalent bond energy by back-calculation. However, results agree reasonably well with the value of 43.4 kpm, which is based on an electronegativity value revised somewhat upward from the earlier 2.84 to 2.98, and calculated from the $E = CrS$ equation where $C = 11.2$ corresponding to $n = 4$.

Gallium

The nonpolar covalent radius of gallium extrapolated from iodine and gray tin using the Z/r^3 function is 1.26 Å and the electronegativity is taken as 3.28. From the $E = CrS$ equation the homonuclear single covalent bond energy is calculated as 46.3 kpm. This is tentatively adopted.

Thermochemical data for gallium compounds are inadequate for reliable estimation of the bond energy by back-calculation. The dissociation energy of $Ga_2(g)$ is calculated to be 27.6 kpm from the enthalpies of formation of 66.2 for Ga and 104.8 for Ga_2. The lack of agreement is duly noted but as yet there is insufficient evidence for judgment.

Germanium

The experimental bond length in germanium is 2.44 Å corresponding to a nonpolar covalent radius of 1.22 Å. The electronegativity is 3.59. The experimental atomization energy of germanium is 90.0 kpm. Since atomization of its diamond structure would involve the equivalent of breaking two Ge—Ge bonds per atom, a homonuclear single covalent bond energy of 45.0 kpm would seem appropriate, assuming the discrepancies noted for carbon and perhaps silicon to be absent here. However, the value calculated from the $E = CrS$ equation is 49.1, which is appreciably higher. Back-calculation from the experimental atomization energies of the gaseous halides gives the following values: $GeCl_4$, 48.5; $GeBr_4$, 52.0; GeI_4, 49.2; averaging 49.9 (all based on the X''' energies of the halogens). Therefore the value 49.1 is tentatively adopted.

Arsenic

The bond length in As_4 is 2.43 corresponding to a nonpolar covalent radius of 1.22 Å. However, a "tetrahedral" radius of 1.18 is reported, and interpolation of the Z/r^3 function between germanium and bromine gives 1.19 Å, which is the value adopted here.

The electronegativity of arsenic is 3.90. The experimental heats of formation are 72.3 for monatomic $As(g)$ and 34.4 for $As_4(g)$ from which the atomization energy of As_4 is 254.8 kpm. Atomization of As_4 requires breaking six bonds per molecule, giving an As—As bond energy of 42.5 kpm. However, the bond length is 2.43 Å instead of the covalent radius sum of 2.38 Å. Correction to the latter gives 43.4 for the As' homonuclear single covalent bond energy.

The As''' bond energy calculated from the $E = CrS$ equation is 52.0 kpm. Halfway between 52.0 and 43.4 is 47.7 kpm, the value of the As'' energy.

Selenium

The bond length in elemental selenium is 2.32 Å corresponding to a nonpolar covalent radius of 1.16 Å, which is also the value interpolated between germanium and bromine using the Z/r^3 function. The electronegativity is 4.21. Although there is no known application at present, the calculated homonuclear single covalent bond energy for Se''' from the $E = CrS$ equation is 54.6 kpm. The Se' energy has been reported as about 44 kpm. This would correspond to a Se'' energy of 49.3 kpm.

It is interesting that the standard heat of formation of $Se(g)$ is 49.2 kpm. If atomization of selenium from its standard state is equivalent to the breaking of one covalent bond per atom, this suggests that in selenium as in sulfur (S_8) the Se'' bond energy applies. Further, the heat of formation of $Se_2(g)$ is 34.9, from which the energy of the Se—Se bond in this molecule is 63.5 kpm. The bond length is 2.15 Å, compared to the covalent radius sum of 2.32 Å. If 49.2 is the correct energy for Se'', then the bond in Se_2 cannot possibly be a double bond. The calculated multiplicity factor is 1.20 instead of the olefinic factor of 1.50. Data are lacking for testing the validity of these energies.

Bromine

The bond length in Br_2 is 2.28 Å from which the nonpolar covalent radius of 1.14 Å is obtained. The relative compactness electronegativity is 4.53. The experimental dissociation energy of Br_2 is 46.1 kpm, which is adopted for the Br' homonuclear single covalent bond energy.

Application of the $E = CrS$ equation gives a value of 58.0 for the Br''' parameter. The Br'' energy is then 52.0 kpm. The 58.0 value is supported by back-calculation from the atomization energies of a number of gaseous bromides, giving results as follows: $BeBr_2$, 54.6; BBr_3, 55.0; $AlBr_3$, 61.1; $GeBr_4$, 64.0; and $SnBr_4$, 53.7. The average is 57.5 kpm.

Rubidium

The diatomic molecules of rubidium that exist in small concentration in rubidium vapor have a bond length of 4.32 Å corresponding to a nonpolar covalent radius of 2.16 Å. The electronegativity is 0.36. The experimental dissociation energy of Rb_2 at 25° is 12.4 kpm. The homonuclear single covalent bond energy calculated by the equation $E = 16.1 \, rS$ 12.4 kpm.

Strontium

The nonpolar covalent radius of strontium is taken as 1.91 Å and the electronegativity as 1.06. By the $E = CrS$ equation, the homonuclear single covalent bond energy is calculated as 32.6 kpm. No direct experimental verification is possible, but back-calculations can be made from the atomization energies of the solid halides. Results are: SrF_2, 32.1; $SrBr_2$, 32.6; SrI_2, 33.4; average, 32.7 kpm. Therefore the value 32.6 is tentatively adopted.

Cadmium

The nonpolar covalent radius of cadmium extrapolated from iodine and tin using the Z/r^3 function is 1.46 Å. The electronegativity is taken as 2.59. From the $E = CrS$ equation, where $C = 8.6$ corresponding to $n = 5$, the homonuclear single covalent bond energy of cadmium is calculated to be 32.5 kpm. These values give only rough agreement between calculated and experimental atomization energies for the limited number of cadmium data available. However, agreement would be much improved if the atomization energy of cadmium were 42.7 instead of 27. Problems obviously exist here which have not yet been solved.

Indium

The nonpolar covalent radius of indium also extrapolated from iodine and tin is 1.43 Å. The electronegativity is 2.84. From the $E = CrS$ equation, the homonuclear single covalent bond energy of indium is calculated to be 35.0 kpm. An attempt to verify this value by calculation of the atomization energies of the gaseous halides gives results that are about 13 kpm too low for chloride, bromide, and iodide. This suggests that if the atomization energy of indium were about 45 kpm instead of the reported 58, excellent agreement would be observed between the experimental and calculated atomization energies.

A value of only 25.3 kpm is calculated from the experimental heats of formation of 58.2 for In(g) and 91.0 for $In_2(g)$.

Tin

The experimental bond length in gray tin is 1.40 Å. The electronegativity is 3.09. The homonuclear single covalent bond energy calculated from these data and the $E = CrS$ equation is 37.2. The experimental atomization energy of gray tin is 72.2 kpm corresponding to a bond energy of 36.1 kpm. Too few data are available to justify a selection. Either value gives satisfactory agreement between experimental and calculated bond energies in tin compounds.

In the +2 state the presence of the inert pair is believed to reduce the electronegativity of tin to a value of about 2.31. The above data therefore probably apply only to tetravalent tin.

Antimony

The nonpolar covalent radius of antimony interpolated from the Z/r^3 function of tin and iodine is 1.38 Å. The electronegativity is 3.34. From the $E = CrS$ equation, the homonuclear single covalent bond energy is calculated to be 39.6 kpm. This would be the Sb''' value. An Sb' value of 32.0 has been reported. The Sb'' value would then be 35.8 kpm.

Tellurium

In literature compilations, the nonpolar covalent radius of tellurium has been reported as 1.37 for the "normal" radius and 1.32 for the "tetrahedral" radius. By interpolation between tin and iodine, the value 1.35 Å is obtained. The electronegativity of tellurium is 3.59. By use of the $E = CrS$ equation, the homonuclear single covalent bond energy for Te''' is calculated as 41.7 kpm. The Te' energy has been reported as about 34 kpm, which would make the Te'' energy about 38 kpm.

Iodine

The bond length in I_2 molecule is 2.66 Å from which the nonpolar covalent radius is 1.33 Å. The relative compactness electronegativity is 3.84. The dissociation energy of I_2 is 36.1 kpm, which is the value for the homonuclear single covalent bond energy for I'. The value for I''' is calculated from the $E = CrS$ equation to be equal to 43.9 kpm, which would correspond to an I'' value of 40.0

kpm. Fewer examples of its application exist, but the I''' value is supported by back-calculations from gaseous BeI_2 (43.2) and AlI_3 (42.1 kpm).

Cesium

The bond length in the diatomic molecule of $Cs_2(g)$ is 4.70 Å corresponding to a nonpolar covalent radius of 2.35 Å. The electronegativity is 0.28. The dissociation energy of $Cs_2(g)$ is experimentally determined as 10.7 kpm. The $E = CrS$ equation using the above data gives 10.6 kpm.

Barium

The nonpolar covalent radius of barium is taken as 1.98 Å and the electronegativity 0.78. Application of the $E = CrS$ equation gives 24.9 kpm for the homonuclear single covalent bond energy. An empirical relationship between radii and bonding energies in diatomic molecules and body-centered cubic metallic lattice described in the following chapter gives for barium the homonuclear single bond energy of 25.7 kpm. These values give good agreement between experimental and calculated atomization energies of barium compounds.

OTHER MAJOR GROUP ELEMENTS

For a number of other elements which are outside the present scope of the $E = CrS$ relationship, it is possible to make reasonable estimates of their homonuclear single covalent bond energy by back-calculation from experimental atomization energies. This procedure without outside support has, of course, a circular character that may reasonably impair confidence, since it involves using experimental bond energies to calculate energies which are used only for calculating bond energies. Nevertheless, it is significant that essentially the same values are calculated from a series of different compounds. The following data are tentative, and their basis is fully described to permit individual judgment as to their possible validity.

Mercury

Uncertainty as to structural parameters of solid mercury compounds requires reliance on data for the gaseous dihalides to estimate the homonuclear single covalent bond energy of mercury. These data are quite inadequate. In $HgF_2(g)$, the covalent energy contribution is too small for use in back-calculation of bond energy. The electronegativity of mercury is 2.93 and the nonpolar covalent radius 1.49 Å. If the halogen is assumed to use its single prime energy in its bonds to mercury, then the following Hg—Hg energies are back-calculated: $HgCl_2$, 7.5; $HgBr_2$, 7.1; HgI_2, 11.3; average, 8.6 kpm. Use of this average to calculate atomization energies gives the following results (calculated—experimental): HgF_2, 145.0–122.7; $HgCl_2$, 110.4–107.9; $HgBr_2$, 91.8–88.5; HgI_2, 64.8–69.6 kpm.

Thallium

Data are available only for the (I) state of thallium—the halides. The electronegativity of thallium (I) is 1.89 and its covalent radius is 1.48 Å. From the four solid halides can be back-calculated an average Tl—Tl energy of 17.8 kpm. Use of this value in calculating the atomization energies of the halide salts gives the following results (calculated—experimental): TlF, 141.5—140.1; TlCl, 122.0—121.7; TlBr, 111.4—111.5; TlI, 97.1—98.8 kpm.

Lead

The electronegativity of lead (II) is 2.38 and that of lead (IV) 3.08. The covalent radius is 1.47 Å. Numerous data for lead halides are available. Back-calculation of the Pb—Pb energy from the solid halides gives the following results: PbF_2, 31.2; $PbCl_2$, 17.4; $PbBr_2$, 20.3; PbI_2, 24.1 kpm. The gaseous mono- and dihalides appear to involve the triple prime energies of the halogen atoms, which by back-calculation give the following Pb—Pb energies: Pb—Cl, 23.4; $PbCl_2$, 19.7; PbBr, 19.7; $PbBr_2$, 19.9; PbI, 19.0; PbI_2, 20.7 kpm. The value for $PbF_2(c)$ was omitted in averaging to give the value 20.5. This is the adopted value.

The atomization energy of $PbF_4(g)$ is clearly too low to involve the F''' energy. It appears that the X' energy is applicable to the tetrahalides. Back-calculation of the Pb—Pb energy from these gives the following results: $PbCl_4$, 20.6; $PbBr_4$, 18.8; and PbI_4, 21.4; average 20.3, practically the same as for the Pb(II) compounds. It is therefore assumed correct and that the Pb—Pb energy does not depend on the oxidation state or number of bonds formed by the lead. This is of course consistent with earlier results.

Bismuth

Bismuth has an electronegativity value of 3.16 and a covalent radius of 1.46 Å. Data on arsenic and antimony halides, although generally poor, suggest that the triple prime energy of the halogen should be applicable to the bismuth trihalides. Back-calculating from experimental atomization energies leads to an average value of 30.4 kpm for the homonuclear single covalent bond energy of bismuth. Application of this value to the three trihalides for which data are available gives the following results (calculated—experimental): $BiCl_3$, 201.3—200.0; $BiBr_3$, 167.7—168.1; and BiI_3, 129.6—129.9 kpm.

TRANSITIONAL ELEMENTS

Bond energy calculations for the transitional elements are complicated by uncertainties as to crystal field effects, electronegativities, covalent radii, and other factors. However, $3d$ electron effects can be considered minimized in compounds of titanium (IV), manganese (II), and copper (I) and silver (I).

Titanium

Data for the gaseous tetrahalides are available. The F'' energy appears appropriate for TiF_4 but the X' energy for the other halides. From these a Ti—Ti energy is back-calculated with the following results: $TiCl_4$, 13.0; $TiBr_4$, 11.1; TiI_4, 17.3; average 13.8 which is adopted. When this value is applied to calculating the atomization energy of titanium (IV) compounds, the following data are obtained for the halides (calculated—experimental atomization energies): TiF_4'', 558.4—558.8; $TiCl_4$, 412.0—411.3; $TiBr_4$, 355.2—350.0; TiI_4, 277.2—283.0. The value can also be applied to solid TiO_2, giving a calculated atomization energy of 459.6 to be compared with the experimental value of 458.0 kpm.

The covalent radius is 1.32 Å and the electronegativity 1.40.

Manganese

The covalent radius is 1.17 Å and the electronegativity 2.07. Back-calculation of the Mn—Mn energy from atomization energies of the solid compounds gives the following results: MnF_2, 31.9; $MnCl_2$, 30.6; $MnBr_2$, 30.7; MnI_2, 35.7; MnO, 39.6; MnS, 31.6; average value of 33.4 kpm is adopted.

Copper

The nonpolar covalent radius of copper extrapolated by the Z/r^3 function from germanium and bromine is 1.35 Å. The electronegativity is 2.60. Back-calculation from atomization energies experimentally determined for the solid (I) halides gives the following values of the Cu—Cu energy: CuF, 15.6; CuCl, 15.5; CuBr, 16.9; CuI, 20.2; average value of 17.1 is adopted.

Silver

The nonpolar covalent radius of silver extrapolated by the Z/r^3 function from tin and iodine is 1.50 Å. The electronegativity is 2.57. Back-calculation from the experimental atomization energies of the solid halides gives the following values of the Ag—Ag energy: AgF, 17.6; AgCl, 14.1; AgBr, 15.8; AgI, 15.2; the average of 15.7 adopted.

SUMMARY

The numerical data for this chapter are summarized in Table 3-1, together with other necessary or helpful data for bond energy calculations.

From the above details it should be clear that experimental data have been used wherever possible to validate or provide the selection of fundamental atomic properties to be used for the calculation of bond energies. Although in some cases

TABLE 3-1

Selected Values of Atomic Properties

Element	r_c(A)	S	E'''	E''	E'	$\Delta Hf°$ (g) (kpm)
H	0.32	3.55	104.2	–	–	52.1
Li	1.34	0.74	28.3	–	–	38.4
Be	0.91	1.99	51.6	–	–	78.3
B	0.82	2.93	68.4	–	–	134.5
C	0.77	3.79	83.2	–	–	171.3
N	0.74	4.49	94.8	67.0	39.2	113.0
O	0.70	5.21	103.9	68.7	33.5	59.6
F	0.68	5.75	111.4	75.5	39.5	18.9
Na	1.54	0.70	18.0	–	–	25.8
Mg	1.38	1.56	34.6	–	–	(47)
Al	1.26	2.22	45.0	–	–	78.0
Si	1.17	2.84	53.4	–	–	108.9
P	1.10	3.43	60.7	55.9	51.1	79.8
S	1.04	4.12	69.0	62.1	55.0	66.6
Cl	0.99	4.93	78.6	68.4	58.2	29.1
K	1.96	0.42	13.2	–	–	21.3
Ca	1.74	1.22	34.2	–	–	42.5
Zn	1.30	2.98	43.4	–	–	31.
Ga	1.26	3.28	46.3	–	–	66.2
Ge	1.22	3.59	49.1	–	–	90.0
As	1.19	3.90	52.0	47.7	43.4	72.3
Se	1.16	4.21	54.6	49.3	44.0	49.2
Br	1.14	4.53	58.0	52.0	46.1	26.7
Rb	2.16	0.36	12.4	–	–	19.6
Sr	1.91	1.06	32.6	–	–	39.
Cd	1.46	2.59	32.5	–	–	27.
In	1.43	2.84	35.0	–	–	(45.2)
Sn(II)	1.40	2.31	37.2	–	–	72.2
Sn(IV)	1.40	3.09	37.2	–	–	72.2
Sb	1.38	3.34	39.6	–	–	62.7
Te	1.35	3.59	41.7	38.	34.	45.5
I	1.33	3.84	43.9	40.0	36.1	25.5
Cs	2.35	0.28	10.7	–	–	18.7
Ba	1.98	0.78	24.9	–	–	42.
Hg	1.49	2.93	8.6	–	–	14.7
Tl(I)	1.48	1.89	17.8	–	–	43.6
Pb(II)	1.47	2.38	20.5	–	–	46.8
Pb(IV)	1.47	3.08	20.5	–	–	46.8
Bi	1.46	3.16	30.4	–	–	49.5
Ti(IV)	1.32	1.40	13.8	–	–	113.0
Mn(II)	1.17	2.07	33.4	–	–	67.
Cu(I)	1.35	2.60	17.1	–	–	81.
Ag(I)	1.50	2.57	15.7	–	–	68.

the choice of atomic property was somewhat arbitrary, in general they are well founded and should be reasonably reliable. The interrelationship between electro-negativity and homonuclear single covalent bond energy is sufficiently quantitative to justify the faith that both quantities are based on the same nuclear—electronic interactions, and for most of the elements are experimentally verifiable through quantitative measurement of the homonuclear bond energy.

It is only fair to point out that even if the electronegativity and bond energy were pulled out of a hat, their quantitative applicability to the heretofore very difficult problem of calculating bond energies in heteronuclear bonds would be extremely impressive. The data of the following chapters should supply reinforce-ment to all the concepts on which the calculations are based, as well as to the fundamental properties of Table 3-1.

FOUR

The Physical States
of Nonmetals

THE MONATOMIC GASEOUS ELEMENTS OF GROUP M8

The helium family of chemical elements, in which each element beyond helium has eight outermost electrons, is called Group M8 according to the proposal[1] that ambiguity in group numbering be removed by labeling major group or "main" group elements M and transitional elements T. These are the only elements for which the normal physical state at ordinary temperatures is monatomic molecular gaseous. Contrary to common explanation, there is nothing magical about these elements and their particular numbers of electrons except that in this arrangement the nucleus is so effectively shielded by the electrons that no orbital vacancies persist in which the nuclear charge can be effectively sensed. This fact is quite sufficient, however, to prevent any covalent bonding between like atoms of M8 elements and to account fully for their normal existence as monatomic gases. They are not ideal gases, of course, although more nearly so than other substances. All give evidence of appreciable van der Waals interactions when sufficiently cooled, and only helium requires external pressure as well as cooling to transform it from liquid to solid.

Any book on chemical bonding must take appropriate note of the lack of bonding in these elements and the presence of bonding in all other elements. But the reader must be warned to avoid the historical trap of believing there is something miraculous about the particular numbers of electrons in the clouds of

[1] R. T. Sanderson, *J. Chem. Educ.* **41**, 187 (1964).

the atoms of these elements, 2, 8, 18, 36, 54, and 86. In fact, the "miracle" exists **only when those particular numbers are accompanied by** *equal* **nuclear charges.**

To appreciate this, imagine being able to make changes in the nucleus without touching the electronic cloud. For example, picture an atom of argon, with its nuclear charge of +18 imbedded in its cloud of 18 electrons. Removal of one proton would upset the balance between cloud compactness and nuclear charge. Now less restrained by the +17 nucleus, the cloud must expand to a larger volume. But most important, the argon atom is thus transmuted to a chloride ion, Cl^-. A chloride ion, of course, in no way, *except* for electron population, resembles an argon atom. A chloride ion can be oxidized, it can act as electron pair donor in the formation of complexes, it can be joined by oxygen atoms to form various familiar anions, it can join all manner of cations to form salts (chlorides)—none of which is possible for argon atoms. The reason that chlorine tends to acquire one electron, then, is certainly not, as we are led from chemical infancy to suppose, to "become like" argon. A chlorine atom tends to acquire one electron because it has one outermost vacancy capable of accommodating such an electron together with a substantially high effective nuclear charge to attract that electron. The reason the chlorine atom tends to acquire *only* one electron is primarily that it has only one vacancy. If an argon atom with an 18+ nuclear charge fails to attract a foreign electron, then certainly a chloride ion with a 17+ charge at the nucleus cannot be expected to.

Similarly, one might add one proton to the argon nucleus. This too would upset the balance between nucleus and cloud, now causing the cloud to be pulled in more tightly since the nuclear charge is greater. In fact, by this little deed, a potassium ion, K^+, is created. A potassium ion can be reduced, it can become hydrated, it can become coordinated with all manner of negatively charged ions to form potassium salts—and, of course, none of these is possible for argon. The reason an atom of potassium tends to lose one electron is not because of some inherent urge to become like argon, but rather because potassium holds its outermost electron relatively weakly and because practically every other active element attracts electrons more strongly than potassium and can overcome in a competition. The reason an atom of potassium tends to lose only one electron is not that it has achieved the magic electron number of 18, but that it would require so much greater energy to remove an electron from the underlying shell of eight that no chemical reaction can supply it.

In summary, reform in the practice of presenting a fictional account of chemical bonding to beginning students is long overdue. The M8 elements are the cornerstone of the periodic system not because they represent some marvelous ideal that every other element strives to emulate or attain. They represent the end of each period, where one more unit increase in atomic number will initiate a new outermost principal quantum level and therefore begin a new period. That is their importance and their significance. In addition, they provide the fundamental clue to possibilities of covalence. Outermost and reasonably stable vacancies are an absolute essential.

THE HALOGENS

Atoms of all other elements outside of Group M8 have outermost vacancies within which the effective nuclear charge of the atom can be sensed. The halogens, Group M7, consist of atoms having just one outermost vacancy, along with seven outer electrons. Thus each atom has an outermost half-filled orbital and the capacity to form a single covalent bond. Consequently, when these atoms come together, diatomic molecules held together by a single covalent bond each are formed. This completely utilizes the bonding capacity of the halogen atoms, so that they are now unable to undergo further reaction except by breaking apart into atoms again.

The strength of the bonds so formed must of course depend on the atomic radius and the effective nuclear charge. This strength can be measured accurately experimentally as the dissociation energy of the diatomic molecule. The values for chlorine, bromine, and iodine were known for a long while before it was experimentally possible to evaluate the dissociation energy of fluorine. Values of 58 for chlorine, 46 for bromine, and 36 for iodine suggested, in view of other comparative properties of the halogens, that the dissociation energy of fluorine should be much larger than 58 kpm. As explained in Chapter Two, the experimental value of about 38 kpm was therefore considered anomalous and mysterious until recognized as the consequence of "lone pair weakening." A thorough understanding must await more insight as to the nature of this weakening effect.

The diatomic molecular nature of the halogens does not ensure their existence as gases under normal conditions. Although F_2 melts and boils at very low temperatures, being similar in this respect to O_2, chlorine is rather easily condensed under a pressure of about 10 atmospheres and boils at $-34°C$. Bromine is liquid at ordinary temperatures, boiling at $58.2°C$ and melting at $-7.3°C$. Iodine sublimes readily, melting at $113.6°$ and boiling at $184.5°C$. These differences are attributed to van der Waals forces, which increase substantially with increasing number of electrons: 18, 34, 70, and 106. In the solid state, they all form molecular crystals.

THE CHALCOGENS

The M8 elements all exist as monatomic gases because they have no ability to become bonded to one another, and the M7 elements exist as diatomic molecules because when their univalence is satisfied, no further possibilities of stable bonding exist. With M6 elements, however, each atom has two options from the viewpoint of condensation to polyatomic aggregates. That is, each atom has two half-filled outermost orbitals, giving it the capability of forming two single covalent bonds or one double covalent bond. Here is where the lone pair bond weakening effect becomes especially important, for it appears to determine which will be the characteristic form of the element under normal conditions.

Where the lone pair weakening effect is very large, as in oxygen, then a double bond is favored over two single bonds because in the double bond the weakening effect is partly removed. But where the lone pair effect is much smaller, then the advantages of reduced weakening in multiple bonding are not sufficient, and two single bonds are nevertheless stronger than one double bond. Consequently, sulfur, selenium, and tellurium are polymeric rather than diatomic, as will be detailed presently.

The atomization of polymeric oxygen would require breaking one single bond per mole of oxygen atoms and therefore an energy of 33.5 kpm. It takes 59.6 kpm of oxygen atoms to atomize O_2 molecules. Clearly the diatomic molecule represents a much more stable aggregate than the polymer, depolymerization being expected with both evolution of heat and increase in entropy.

The only even moderately successful extension of a chain of oxygen atoms beyond two is shown by ozone, in which three oxygen atoms are so linked. Here the central atom is incapable of forming two double bonds at once, the closest approach apparently being for each bond to average single and double. The single bond of necessity is a coordination bond with terminal oxygen as acceptor. The energies involved and the consequences thereof are discussed on page 132.

It is commonly implied, if not stated, that the heavier atoms are incapable of stable multiple bonding and therefore condense by single bonds, forming polymeric solids. In fact, very stable diatomic molecules form at high temperatures. The principle species in sulfur vapor between 500 and 1900°C is the S_2 molecule. Such molecules appear to resemble O_2 molecules closely. The bond length is considerably shorter than an ordinary single bond, and the molecule is paramagnetic, indicating the presence of two unpaired electrons. The experimental bond energy is 102.5 kpm. From this and the bond length of 1.89 Å, one can calculate an S'' single bond energy of 62.1 kpm. This in itself is too near the S' energy of 55.0 kpm to allow the favoring of a double bond over single bonds, and two single bonds are indeed stronger than a double bond, 110 to 103 kpm. The entropy would decrease with polymerization, diminishing its negative free energy and perhaps helping the formation of dimer. The evidence thus far stated is thus inadequate to convince us that solid polymer should be the preferred form at ordinary temperatures.

However, there is a peculiarity about solid sulfur that enters into the picture. Instead of preferring long chain molecules, as are present in "plastic" sulfur, this element tends to form stable eight-membered rings, S_8, which then are joined by van der Waals forces into the familiar rhombic and monoclinic crystalline forms of sulfur. No one seems to understand why S_8 is favored over other possible rings and chains, but when the atomization energy is determined, it is found to be 66.6 kcal/mole of atoms. From this and the energy of vaporization of S_8 crystal, one calculates an average bond energy in S_8 of 63.5 kpm, very close to the 62.1 value of E''. It may be assumed, therefore, that there must be something special about S_8 molecules that causes the lone pair weakening effect to be reduced for that particular staggered ring geometry. This quality, whatever it may be, appears to add enough to the stability of solid sulfur to cause it to be favored easily over the gaseous dimer.

Comparable data for selenium and tellurium are sparse, but we may assume some similarity to sulfur.

ELEMENTS OF GROUP M5

The existence of nitrogen as diatomic molecules, N_2, held together by a triple bond has been explained in Chapter Two. One can easily determine that neither chain polymers or ring polymers or "nitrogen benzene" could compete with N_2 molecules because of the large reduction in lone pair bond weakening when a triple bond forms.

The normal vapor species from phosphorus is P_4, in which single bonds exist, but these molecules break into P_2 molecules when heated above 800°C. P_2 molecules in turn are stable over a wide temperature range. The bond length is 1.89 compared to 2.20 Å for the nonpolar covalent radius sum, and the experimental dissociation energy is 125.1 kpm. These suggest that a triple bond holds the two phosphorus atoms together, and that P_2 therefore closely resembles N_2. From these data we can calculate an E''' value of 60.7 kpm, which is exactly the same as the extrapolated value obtained by use of the $E = CrS$ relationship. The bonds in P_2 molecules thus do resemble those in N_2 molecules. There is no basis for assuming that the existence of phosphorus as a solid under ordinary conditions reflects an inability of phosphorus atoms to form stable multiple bonds.

The difference between nitrogen and phosphorus is that for nitrogen the "lone pair bond weakening effect" is very large, whereas in phosphorus it is relatively small—the difference between 60.7 and 51.1 kpm, the latter being the single bond energy where weakening is complete. To atomize P_4, three bonds must be broken per phosphorus atom, but each break liberates two atoms. Therefore the atomization energy is $1.5 \times 51.1 = 76.7$ kcal/mole of atoms. Atomization of P_2, however, would require only half of 125.1, or 62.6 kpm of atoms. There is thus a 14 kpm advantage of P_4 with single bonds over P_2 with triple bond. This is why phosphorus is not a gas like nitrogen, under ordinary conditions.

White phosphorus, P_4, is of course less stable than the other forms because the structure implies 60° angles between bonds, which crowd them. Another approximately 4 kpm can be gained in stability by opening out the bond angles to better than 100°, as in red and other modifications of this element.

Presumably the heavier members of this group are solids for similar reasons.

ELEMENTS OF GROUP M4

Within this group, each atom has four outermost electrons and therefore can provide four outermost half-filled orbitals for the formation of four single covalent bonds. The network solid exemplified by diamond, silicon, and germanium is the natural consequence of this kind of bonding. There is no "lone pair effect" in these elements and therefore little problem of interpretation of structure.

The one problem that does occur is that of carbon in the structure called graphite. Here each carbon atom is connected to three other carbon atoms in the same plane at the corners of an equilateral triangle, suggesting that each bond has a bond order of 1.33. The shorter bond length and higher bond energy are both consistent with this picture. However, the atomization energy of diamond is only about 0.4 kpm lower than that for graphite, and this difference does not seem nearly sufficient to cause carbon atoms to assemble, under normal conditions, practically exclusively in the graphite form rather than diamond.

The following speculative explanation may be offered. A molecule or molecular fragment of formula C_2 seems to have both bond length and bond energy that suggest a double bond. In other words, the C_2 unit is like a molecule of ethylene from which the four hydrogen atoms have been stripped. Ethylene is a planar molecule, and removal of the hydrogen atoms would still leave the four unused bonding orbitals in a planar configuration. The next step in the polymerization of carbon atoms, beginning with the individual atoms and going through the C_2 stage, would be joining together of C_2 fragments into six carbon rings as in graphite. There being then no reason for double bonds to be localized, they would become equalized with the single bonds and the graphite layer would be formed.

In other words, graphite forms because of the nature of the initial C_2 fragments, which cannot assemble to produce the diamond structure without outside assistance of considerable power. In the commercial synthesis of diamond, it would be interesting to know whether pressure application at right angles to the graphite sheets would be more effective at buckling these sheets into the diamond structure than squeezing the layers together. It would also be interesting to know whether it might be possible to begin with a crystal of diamond as a nucleus and build it up with carbon atoms without permitting them first to dimerize.

FIVE

Polar Covalence I:
Electronegativity Equalization,
Partial Charge, and Bond Length

THE PRINCIPLE OF ELECTRONEGATIVITY
EQUALIZATION

The usual definition of electronegativity as "the power of an atom in a molecule to attract electrons to itself" leaves much to be desired in both clarity and significance. Electronegativity as listed is the property of an isolated atom. It changes when the atom is placed in a variety of conditions. A very simple model of the act of formation of a heteronuclear covalent bond can be very helpful in visualizing some of the fundamental consequences of chemical combination.

Let us begin with atoms A and B, B initially more electronegative than A. Let us assume that each atom possesses at least one half-filled outermost orbital, giving it the capacity to form a covalent bond. When the two atoms come in contact, the single electron of A finds the vacancy of B available and the single electron of B finds the vacancy of A available. Thus both electrons come under the considerable influence of both nuclei and are shared between the two atoms. The attraction between the two atoms that results from this mutual sharing of the two bonding electrons is called a covalent bond. Since here the atoms are not alike, the bond is called **heteronuclear**. In the fact of sharing the same two electrons between two nuclei, the heteronuclear bond is just like a homonuclear bond. However, a difference arises from the fact that B is initially more electronegative than A. This implies that the bonding electrons will be more strongly attracted to the nucleus of B than to the nucleus of A.

75

A stable system cannot result unless in effect the bonding electrons are able to adjust to a condition of essentially equal attraction to both nuclei. Electrons are free to move according to the forces acting upon them, and they are certainly not to be expected to remain evenly distributed if they are not evenly attracted. The natural expectation is that the bonding electrons will adjust their average positions in such a way as to become more closely associated with the nucleus of B, and consequently less closely associated with the nucleus of A, than they would be if equal sharing were possible. In fact, one might be inclined to think, at first, that the electrons should move completely to the sphere of influence of that nucleus which attracts them more strongly, abandoning the other nucleus. The fallacy of this expectation will be revealed in just a moment.

Since the atoms were initially electrically neutral, the unevenness of sharing of bonded electrons is certain to upset this neutrality. The atom which gains more than half share, B, acquires a partial negative charge, leaving atom A with a partial positive charge. Why partial charge? Why not a complete transfer? Remember that what attracts an electron to an atom is the effective nuclear charge, which in turn is the net result of the full positive charge on the nucleus and the repulsive effects of the other electrons. The effective nuclear charge is very sensitive to the average electron population around the nucleus. It must change if the electron population changes. As atom B becomes partially negative, this increases the electron population around nucleus B, causing two closely interrelated effects. One is to decrease the effective nuclear charge by increasing the screening by the electrons. The other is to increase the interelectronic repulsions and therefore cause expansion of the electronic cloud. Thus a *reduced* effective nuclear charge is forced to operate over a *greater* distance. This corresponds to a *reduction* in electronegativity.

The opposite situation prevails at atom A. Here the loss of part of the electronic cloud results in reduced shielding of the nucleus and a consequent increase in the effective nuclear charge. The electronic cloud must shrink as the repulsions are diminished and the attractions increased, so the atom grows smaller. A *larger* effective nuclear charge operating over a *shorter* distance corresponds to an *increased* electronegativity. Then why not a *complete* transfer of bonding electrons? Because an *incomplete* transfer is usually enough to reduce the electronegativity of B and increase the electronegativity of A to the extent that they become equal. When both atoms are equal in electronegativity, there is no incentive for further transfer of the bonding electrons.

The principle of electronegativity equalization[1] appears to apply very generally. It may be stated, **"When two or more atoms initially different in electronegativity combine chemically, they become adjusted to the same, intermediate electronegativity within the compound."** This concept remained little noted for a decade or more after publication, following which interest has gradually accumulated.[2] It

[1] R. T. Sanderson, *Science* **114**, 670 (1951).
[2] R. Ferreira, *Trans. Faraday Soc.* **59**, 1064 (1963); J. Hinze, M. A. Whitehead, and H. H. Jaffe, *J. Am. Chem. Soc.* **85**, 148 (1963); G. Klopman, *J. Am. Chem. Soc.* **86**, 1463, 4550

has more recently become an integral part of the calculation of bond energies, as will be demonstrated.

It is of course desirable to know the value of the intermediate electronegativity in a compound. A very satisfactory postulate is that it is **the geometric mean of the individual electronegativities of all the component atoms that make the compound formula.**

PARTIAL CHARGE

A knowledge of the intermediate electronegativity exhibited by the individual atoms in a compound has only limited value, but we have a greater interest in knowing how much partial charge resides on each atom. The principle of electronegativity equalization provides the means for a simple calculation of partial charge. **Partial charge is defined,** for our purposes, **as the ratio of the change in electronegativity undergone by an atom in joining the compound to the change it would have undergone in acquiring or losing one electron.**

Two basic assumptions are needed to evaluate the change in electronegaactivity that would accompany acquisition of unit charge. One is that the partial charge is a linear function of electronegativity. The other is that the bond in a molecule of NaF is 75% ionic. The latter is of course somewhat arbitrary. Originally it was intended only to give reliable relative values of charge. It has been found that the value 75 was a very lucky guess and apparently just about right, so that the charge values obtained have absolute as well as relative significance.

We first determine the electronegativity of NaF as the square root (geometric mean) of the electronegativity of sodium times the electronegativity of fluorine. We then subtract each atomic electronegativity from this molecular electronegativity to determine how much each element has changed. We then divide each change by 0.75, the assumed charge, to calculate the change that would correspond to acquisition of unit charge. Having these values for sodium and for fluorine, we can then perform similar calculations for other compounds of sodium and of fluorine and learn the appropriate values for all the other elements for which electronegativity values are known. A mathematical analysis of this procedure shows that it is equivalent to multiplying the square root of the atomic electronegativity by 2.08, a factor determined by our arbitrary (but correct) choice of 75% ionicity for NaF. In other words, for any element the change in electronegativity that would correspond to acquisition of unit plus or minus charge is $2.08\sqrt{S}$.

(1964); N. C. Baird, J. M. Sichel, and M. A. Whitehead, *Theoret. Chim. Acta* 11, 38 (1968); M. C. Day and J. Selbin, "Theoretical Inorganic Chemistry," 2nd Ed., p. 138, Van Nostrand-Reinhold, Princeton, New Jersey, 1969; "Physical Chemistry," (H. Eyring, D. Henderson, and W. Jost, Eds.), Vol. V, pp. 182, 268, Academic Press, New York, 1970; G. Van Hooydonk and Z. Beckhaut, *Chem. Ber.* 74, 123, 327 (1970); R. S. Evans and J. E. Huheey, *J. Inorg. Nucl. Chem.* 32, 373, 383, 777 (1970).

TABLE 5-1

Data for Calculating Partial Charge on Combined Atoms

Element	Electronegativity[a]			Element	Electronegativity[a]		
	S	$\log S$	ΔS_i		S	$\log S$	ΔS_i
H	3.55	0.5502	3.92	Se	4.21	0.6243	4.27
Li	0.74	−0.1308	1.77	Br	4.53	0.6561	4.43
Be	1.99	0.2989	2.93	Rb	0.36	−0.4437	1.25
B	2.93	0.4669	3.56	Sr	1.06	0.0253	2.14
C	3.79	0.5786	4.05	Cd	2.59	0.4133	3.35
N	4.49	0.6522	4.41	In	2.84	0.4533	3.51
O	5.21	0.7168	4.75	Sn(II)	2.31	0.3636	3.16
F	5.75	0.7597	4.99	Sn(IV)	3.09	0.4900	3.66
Na	0.70	−0.1549	1.74	Sb	3.34	0.5237	3.80
Mg	1.56	0.1931	2.60	Te	3.59	0.5551	3.94
Al	2.22	0.3464	3.10	I	3.84	0.5843	4.08
Si	2.84	0.4533	3.51	Cs	0.28	−0.5528	1.10
P	3.43	0.5353	3.85	Ba	0.78	−0.1079	1.93
S	4.12	0.6149	4.22	Hg(II)	2.93	0.4669	3.59
Cl	4.93	0.6928	4.62	Tl(I)	1.89	0.2765	2.85
K	0.42	−0.3768	1.35	Pb(II)	2.38	0.3766	3.21
Ca	1.22	0.0864	2.30	Pb(IV)	3.08	0.4886	3.69
Zn	2.98	0.4742	3.58	Bi(III)	3.16	0.4997	3.74
Ga	3.28	0.5159	3.77	Ti(IV)	1.40	0.1461	2.48
Ge	3.59	0.5551	3.94	Mn(II)	2.07	0.3160	2.99
As	3.90	0.5911	4.11	Cu(I)	2.60	0.4150	3.36
				Ag(I)	2.57	0.4099	3.33

[a] S is electronegativity. Log S is used to determine geometric mean of S. Partial charge is the ratio of ΔS in the compound to ΔS_i for unit charge.

Data needed and useful for calculating partial charge are presented in Table 5-1. The logarithm of the electronegativity, $\log S$, is of course useful in determining the geometric mean in a compound. For convenience, a short log table is provided at the end of the book, preceding the index.

Partial charge so calculated will be demonstrated to be absolutely essential, and also correct, for the calculation of the energies of heteronuclear covalent bonds. As will be shown in the following section of this chapter, the partial charge can be very usefully employed in estimating the bond length for certain types of compounds. In addition, the partial charge on a combined atom is an extremely useful index of the condition of that atom in the compound. Both physical and chemical properties of compounds can be correlated with the partial charges on their component atoms.[3] Some examples of this kind of application, as well as the

[3] R. T. Sanderson, "Chemical Periodicity," Van Nostrand-Reinhold, Princeton, New Jersey, 1960; "Inorganic Chemistry," Van Nostrand-Reinhold, Princeton, New Jersey, 1967.

applications to bond energies, will also be provided in later chapters. For the present, it may be interesting to outline the changes in atomic properties that accompany partial charge acquisition. This should indicate in a general way the extensive and fascinating possibilities.

When neutral a given atom will have certain characteristics that influence its chemistry, such as oxidizing power, reducing power, electron donating power, electron pair accepting power, and so on. To the extent that this atom acquires partial negative charge in a compound, it must become less oxidizing, more reducing, less able to accept an electron pair, and more generous in coordinating its outer electron pairs with other atoms. To the extent that the atom acquires partial positive charge in the compound, it must increase its oxidizing power, become less reducing, tend toward improved electron pair attraction in coordination, and decrease its ability to coordinate through electron pair donation. These differences are also distinguishable in physical properties.

Like electronegativity itself, partial charge is a property whose value must be established by practical application. Atomic charges calculated through wave mechanical methods appear sometimes similar to, sometimes different from, those used herein. In some instances the more sophisticated methods produce values that are chemically absurd. A number of experimental measurements intended to enlighten regarding the electronic environment of an atomic nucleus have been interpreted as indicating relative partial charge. Perhaps the most promising at present is X-ray photoelectron spectroscopy (ESCA), the results of which a number of recent workers have tried to correlate with atomic charges. Some of the most recent work, by Hercules and co-workers,[4] has favored partial charges such as described herein as being superior to values obtained by much more sophisticated and difficult quantum mechanical procedures. The overall evidence and consistency of partial charges from electronegativity equalization seem undeniable. In addition to the work just cited, some of this support comes from estimates of atomic radii and bond lengths.

THE RADII OF PARTIALLY CHARGED ATOMS, AND THEIR SUM AS BOND LENGTH

The equilibrium distance between the nuclei of two atoms joined by a chemical bond in a molecule or a nonmolecular solid is called the bond length. The most obvious, and seemingly the most possible, means of calculating a bond length is by assuming it to be the sum of contributions from each atom and attempting to assess each individual contribution as the radius of the atom in the direction of the bond. The alternative, of determining the equilibrium separation of the two nuclei

---

[4] J. C. Carver, R. C. Gray, and D. M. Hercules, *J. Am. Chem. Soc.* **96**, 6851 (1974); W. E. Swartz Jr., R. C. Gray, J. C. Carver, R. C. Taylor, and D. M. Hercules, *Spectrochimica Acta* **30A**, 1561 (1974).

within the total electronic cloud and nuclear arrangement of a molecule or non-molecular crystal, has seemed far too complex a problem to find a practical, general solution. Consequently, much effort has been expended toward assigning radii to the participating atoms such that their sum represents the bond length accurately. It was hoped, of course, that a given atom might exhibit a consistent radius in different combinations. This hope has been largely abandoned with respect to molecules, except for only slightly polar bonds, but it still persists inherently in the "ionic" model of nonmolecular solids. Previous work has been very adequately reviewed, as in various well-known books.[5,6]

All methods of evaluating "covalent" and "ionic" radii of atoms have agreed that the radii of anions are always significantly larger than radii of neutral atoms and that radii of cations are always significantly smaller than radii of neutral atoms. It is reasonable to suppose that the transition from covalent to ionic condition should be accompanied, therefore, by a steady change in atomic radius, corresponding to the change in effective nuclear charge that must also occur.

This is not a new idea. It was applied early to the concept of electronegativity equalization, as previously discussed. Although this work at that time had only limited success in accounting for bond lengths, it did lead to the modified concept of polar covalence described at the beginning of this chapter. The physical picture of electronegativity equalization calls for uneven sharing of the bonding electrons. This imparts a partial negative charge to the initially more electronegative atom and leaves a partial positive charge on the initially less electronegative atom. The negative charge corresponds to expansion of the electronic cloud, increasing the effective radius of the atom, while the positive charge corresponds to a contraction of the electronic cloud, decreasing the atomic radius. The length of a polar covalent bond should therefore be representable as the sum of the two atomic radii. What is needed is some simple basis for evaluating the radius of a charged atom.

Of several possible functions, I have selected the simplest, according to which the radius of a combined atom may be represented as a linear function of the charge:

$$r = r_c - B\delta$$

The radius r of the combined atom equals the nonpolar covalent radius r_c minus the product of the empirical factor B and δ, the partial charge on the atom, calculated as earlier described.

Since the correct value of the bond length can be represented by a wide variety of atomic radii, requiring only that their sum equal the bond length, it is essential to establish a base from which to work. In his well-known studies of bond lengths and "ionic" radii in crystalline solids, Pauling suggested that the experi-

[5] L. Pauling, "The Nature of the Chemical Bond," 3rd Ed., Cornell University Press, Ithaca, New York, 1960.

[6] M. C. Day and J. Selbin, "Theoretical Inorganic Chemistry," 2nd Ed., Van Nostrand-Reinhold, Princeton, New Jersey, 1969.

mental internuclear distance in an isoelectronic ionic crystal such as NaF (wherein both "ions" are thought of as controlling 10 electrons) should be apportioned between the two atoms in inverse proportion to their effective nuclear charges. I have adopted this idea for evaluating radii in both gaseous molecule and non-molecular solid, for the latter making the assumption (Chapter Six) that the components of binary solids are better represented as partially charged atoms than as actual ions.

The experimental bond length in gaseous NaF is reported as 1.93 Å. Pauling gives the total screening constant of 4.52 for a system of 10 electrons. This is almost identical with the Slater values of $2 \times 0.85 = 1.70$, plus $8 \times 0.35 = 2.80$, $= 4.50$. Thus the effective nuclear charge is $9 - 4.52 = 4.48$ for fluoride ion and $11 - 4.52 = 6.48$ for sodium ion. However, partial charges of 0.75 on sodium and -0.75 on fluorine are well supported and appear more realistic. Since these correspond to 0.25 electron extra on a sodium ion, and short on a fluoride ion, these effective nuclear charges are corrected by one-fourth the screening value of 0.35, or 0.09. The corrected effective nuclear charges are 6.39 for sodium and 4.57 for fluorine. Corresponding apportionment of the bond length in gaseous NaF gives radii of 0.81 Å for sodium and 1.12 Å for fluorine. Thus the radius of sodium has decreased from its nonpolar covalent value of 1.54 Å by 0.73 Å. The change corresponding to complete loss of an electron, 0.73/0.75, is 0.97 Å, the value of B in the equation above, for sodium in its compounds in the gas phase: $r = 1.54 - 0.97\delta$.

Similarly, the radius of fluorine is calculated to increase from 0.68 Å, the revised nonpolar covalent value, by 0.59, or to 1.27 for the hypothetical gaseous fluoride ion. The value of B is 0.59 for fluorine in gaseous fluorides: $r = 0.68 - 0.59\delta$. Values for the solid state are similarly based on the apportionment of the bond length, 2.31 Å, in solid NaF. By extensions of these basic values to other elements, and by appropriate averaging techniques, the data of Table 5-2 were assembled.

The application of these data to the calculation of bond lengths in binary compounds is summarized in Table 5-3. Bond lengths were calculated for 223 binary compounds. Experimental data for testing the accuracy of the calculations were available for only 203, the other 20 calculated values being predictions. The compounds tabulated include the gaseous and solid halides of the major group M1 and M2 elements and the solid chalcides of the M2 elements. The average difference between calculated and experimental bond length for the 203 binary compounds is 0.023 Å. Since the individual atomic radii are only calculated to the nearest 0.01 Å, the agreement is satisfactory. The differences in fact do not necessarily indicate errors in calculation, for some of the experimental data are not known with great precision, especially for solids.

The complete data are summarized in Tables 5-4 and 5-8 for gaseous and solid fluorides, Tables 5-5 and 5-9 for gaseous and solid chlorides, Tables 5-6 and 5-10 for gaseous and solid bromides, and Tables 5-7 and 5-11 for gaseous and solid iodides. It is tempting to attribute some special absolute significance to the calculated values of the atomic radii, but these values must not be taken too

TABLE 5-2

Radius Factors and Radii of Hypothetical Ions of Unit Charge

Element	Nonpolar covalent radius (AU)	Charge	B^a, gas	r_i, gas	B, solid	r_i, solid
H	0.32	+	0.974	–	–	–
Li	1.34	+	1.201	0.14	0.812	0.53
Be	0.91	+	0.597	0.31	0.330	0.58
B	0.82	+	0.591	0.23	–	–
C	0.77	+	0.486	0.28	–	–
N	0.74	+	0.311	0.43	–	–
N	0.74	–	0.063	0.80	–	–
O	0.70	–	0.240	0.94	0.401	1.10
F	0.68	–	0.536	1.22	0.925	1.61
Na	1.54	+	0.972	0.57	0.763	0.78
Mg	1.38	+	0.638	0.74	0.349	1.03
Al	1.26	+	0.580	0.68	–	–
Si	1.17	+	0.587	0.58	–	–
P	1.10	+	0.404	0.70	–	–
S	1.04	+	0.681	0.36	–	–
S	1.04	–	0.222	1.26	0.657	1.70
Cl	0.99	–	0.727	1.72	1.191	2.18
K	1.96	+	1.097	0.86	0.956	1.00
Ca	1.74	+	0.691	1.05	0.550	1.19
Ti	1.32	+	0.330	0.99	–	–
Ge	1.22	+	0.577	0.64	–	–
As	1.19	+	0.417	0.77	–	–
Se	1.16	+	0.395	0.77	–	–
Se	1.16	–	0.364	1.52	0.665	1.83
Br	1.14	–	0.695	1.84	1.242	2.38
Rb	2.16	+	1.192	0.97	1.039	1.12
Sr	1.91	+	0.649	1.26	0.429	1.48
Ag	1.50	+	–	–	0.208	1.29
Cd	1.46	+	–	–	0.132	1.33
Sn(II)	1.40	+	0.331	1.07	–	–
Sn(IV)	1.40	+	0.542	0.86	–	–
Sb	1.38	+	0.350	1.03	–	–
Te	1.35	–	–	–	0.692	2.04
I	1.33	–	0.705	2.04	1.384	2.71
Cs	2.35	+	1.273	1.08	0.963	1.39
Ba	1.98	+	0.517	1.46	0.348	1.63
Tl	1.48	+	0.701	0.78	–	–
Pb(II)	1.47	+	0.371	1.10	–	–
Pb(IV)	1.47	+	0.301	1.17	–	–
Bi	1.46	+	0.132	1.33	–	–

[a]The radius of an atom in a polar covalent single bond is the nonpolar covalent radius minus B times the charge on the atom: $r = r_c - B\delta$.

TABLE 5-3

Summary of Binary Compounds for which Bond Lengths
Have Been Calculated

Type of compound	Number of compounds		Average difference, calc. and exp. (A)	
	Gas	Solid	Gas	Solid
Fluorides	39−7[a]	12−1	0.03	0.06
Chlorides	36−2	13	0.02	0.02
Bromides	35−4	13	0.01	0.01
Iodides	31−3	13	0.02	0.04
Oxides	−	5	−	0.03
Sulfides	−	5	−	0.03
Selenides	−	5−1	−	0.04
Tellurides	−	5−2	−	0.03
Hydrogen (in addition to HX)	11	−	0.01	−

[a]A second number indicates the compounds for which experimental bond lengths are predicted but no available for comparison. Summary: 223 compounds (and states), 20 predictions; average difference between calculated and experimental bond length, 0.02 A.

literally until more fundamental support can be provided. That the values are reasonably correct may be a fair assumption because of the initial assignment according to effective nuclear charge. Ideally, however, the B factors should be derivable from more fundamental data. This has not yet been achieved. It is also interesting to note that in the solid state both metal and nonmetal atom have larger radii, for the same partial charges, than in the gas phase. This is of course consistent with the concept of polymeric coordination within the solid.

The 18-shell elements, in their gas-phase compounds, appear to provide the only exceptional behavior noted among the binary compounds studied. These elements do not seem to change in radius with charge in a linear manner. As shown in Table 5-12, the bond lengths in these molecules can best be represented as if the nonmetal were normally behaved but the metal atom has a constant radius. No explanation can be offered at this time, except to suggest that relatively slight withdrawal of valence electrons from these atoms produces nearly the maximum contraction of which they are capable, after which they maintain fairly constant radii no matter how positive the charge becomes.

Radius Ratios—Ions versus Charged Atoms

It is interesting to compare these concepts of bonding and bond length with Pauling's treatment of the solid M1 halides. Pauling recognized that radius ratio and

TABLE 5-4

Bond Lengths in Gaseous Fluorides from Radii of
Charged Atoms (AU)

Compound	$-\delta_F$	r_M	r_F	$r_M + r_F$	R_{exp}	R_{diff}
SF_6	0.05	0.84	0.71	1.55	1.56	0.01
SeF_6	0.05	1.05	0.71	1.76	1.70	0.06
SF_4	0.07	0.85	0.72	1.57	1.56	0.01
NF_3	0.07	0.65	0.72	1.37	1.37	0.00
ClF	0.09	0.92	0.73	1.65	1.63	0.02
GeF_4	0.10	0.99	0.73	1.72	1.68	0.04
AsF_3	0.11	1.05	0.74	1.79	1.71	0.08
SF_2	0.12	0.88	0.74	1.62	1.59	0.03
BrF	0.13	1.05	0.75	1.80	1.76	0.04
PbF_4	0.14	1.30	0.76	2.06	2.08	0.02
PF_3	0.14	0.93	0.76	1.69	1.54	0.15
SiF_4	0.15	0.82	0.76	1.58	1.54	0.04
SbF_3	0.15	1.22	0.76	1.98	–	–
BiF_3	0.16	1.40	0.77	2.17	–	–
BF_3	0.18	0.50	0.78	1.28	1.30	0.02
IF	0.21	1.18	0.79	1.97	1.91	0.06
AlF_3	0.24	0.84	0.81	1.65	1.63	0.02
HF	0.25	0.08	0.81	0.89	0.92	0.03
TiF_4	0.28	0.95	0.83	1.78	1.80	0.02
PbF_2	0.29	1.25	0.84	2.09	2.11	0.02
SnF_2	0.30	1.20	0.84	2.04	–	–
BeF_2	0.34	0.50	0.86	1.36	1.43	0.07
MgF_2	0.41	0.86	0.90	1.76	1.77	0.01
CaF_2	0.47	1.09	0.93	2.02	2.02	0.00
TlF	0.49	1.14	0.94	2.08	2.08	0.00
SrF_2	0.50	1.26	0.95	2.21	2.20	0.01
BaF_2	0.57	1.39	0.99	2.38	2.32	0.06
LiF	0.74	0.45	1.08	1.53	1.56	0.03
NaF	0.75	0.81	1.08	1.89	1.93	0.04
KF	0.84	1.04	1.13	2.17	2.17	0.00
RbF	0.86	1.13	1.14	2.27	2.27	0.00
CsF	0.90	1.20	1.16	2.36	2.35	0.01

TABLE 5-5

Bond Lengths in Gaseous Chlorides from Radii of
Charged Atoms (AU)

Compound	$-\delta_{Cl}$	r_M	r_{Cl}	$r_M + r_{Cl}$	R_{exp}	R_{diff}
BrCl	0.04	1.11	1.02	2.13	2.14	0.01
$SeCl_2$	0.05	1.12	1.03	2.15	2.15	0.00
SCl_2	0.06	0.96	1.03	1.99	2.00	0.01
$GeCl_4$	0.07	1.06	1.04	2.10	2.09	0.01
$AsCl_3$	0.07	1.10	1.04	2.14	2.16	0.02
$PbCl_4$	0.09	1.36	1.06	2.42	2.43	0.01
PCl_3	0.09	0.99	1.06	2.05	2.04	0.01
$SnCl_4$	0.10	1.18	1.06	2.24	2.31	0.07
$SbCl_3$	0.10	1.28	1.06	2.34	2.33	0.01
$SiCl_4$	0.11	0.91	1.07	1.98	2.02	0.04
$BiCl_3$	0.11	1.42	1.07	2.49	2.48	0.01
ICl	0.12	1.25	1.08	2.33	2.30	0.03
BCl_3	0.13	0.59	1.08	1.67	1.75	0.08
HCl	0.16	0.16	1.11	1.27	1.27	0.00
$AlCl_3$	0.19	0.93	1.13	2.06	2.06	0.00
$PbCl_2$	0.23	1.30	1.16	2.46	2.46	0.00
$SnCl_2$	0.24	1.24	1.16	2.40	2.42	0.02
$TiCl_4$	0.24	1.00	1.16	2.16	2.19	0.03
$BeCl_2$	0.28	0.58	1.19	1.77	1.77	0.00
$MgCl_2$	0.34	0.95	1.24	2.19	2.18	0.01
$CaCl_2$	0.40	1.19	1.28	2.47	–	–
TlCl	0.40	1.20	1.28	2.48	2.48	0.00
$SrCl_2$	0.43	1.35	1.30	2.65	2.67	0.02
$BaCl_2$	0.49	1.47	1.35	2.82	2.82	0.00
LiCl	0.65	0.56	1.46	2.02	2.02	0.00
NaCl	0.67	0.89	1.48	2.37	2.36	0.01
KCl	0.76	1.13	1.54	2.67	2.67	0.00
RbCl	0.78	1.23	1.56	2.79	2.79	0.00
CsCl	0.81	1.32	1.58	2.90	2.91	0.01

TABLE 5-6

**Bond Lengths in Gaseous Bromides from Radii of
Charged Atoms (AU)**

Compound	$-\delta_{Br}$	r_M	r_{Br}	$r_M + r_{Br}$	R_{exp}	R_{diff}
$SeBr_2$	0.02	1.15	1.15	2.30	2.30	0.00
SBr_2	0.03	1.00	1.16	2.16	2.16	0.00
$AsBr_3$	0.04	1.14	1.17	2.31	2.33	0.02
$GeBr_4$	0.05	1.10	1.17	2.27	2.30	0.03
PBr_3	0.07	1.02	1.19	2.21	2.20	0.01
$SnBr_4$	0.08	1.23	1.20	2.43	2.44	0.01
$SbBr_3$	0.08	1.30	1.20	2.50	2.50	0.00
IBr	0.08	1.27	1.20	2.47	2.45	0.02
$PbBr_4$	0.08	1.37	1.20	2.57	2.58	0.01
$SiBr_4$	0.09	0.96	1.20	2.16	2.15	0.01
$BiBr_3$	0.09	1.42	1.20	2.62	2.63	0.01
BBr_3	0.10	0.64	1.21	1.85	1.87	0.02
HBr	0.12	0.20	1.22	1.42	1.41	0.01
$AlBr_3$	0.17	0.96	1.26	2.22	2.27	0.05
$PbBr_2$	0.20	1.32	1.28	2.60	2.60	0.00
$SnBr_2$	0.21	1.26	1.29	2.55	2.55	0.00
$TiBr_4$	0.21	1.04	1.29	2.33	2.31	0.02
$BeBr_2$	0.25	0.61	1.31	1.92	1.90	0.02
$MgBr_2$	0.31	0.98	1.36	2.34	2.34	0.00
TlBr	0.36	1.23	1.39	2.62	2.62	0.00
$CaBr_2$	0.36	1.24	1.39	2.63	–	–
$SrBr_2$	0.39	1.40	1.41	2.81	2.82	0.01
$BaBr_2$	0.45	1.51	1.45	2.96	2.99	0.03
LiBr	0.61	0.61	1.56	2.17	2.17	0.00
NaBr	0.62	0.94	1.57	2.51	2.50	0.01
KBr	0.71	1.18	1.63	2.81	2.82	0.01
RbBr	0.73	1.29	1.65	2.94	2.94	0.00
CsBr	0.77	1.37	1.68	3.05	3.07	0.02

TABLE 5-7

Bond Lengths in Gaseous Iodides from Radii of
Charged Atoms (AU)

Compound	$-\delta_I$	r_M	r_I	$r_M + r_I$	R_{exp}	R_{diff}
AsI_3	0.00	1.19	1.33	2.52	2.52	0.00
GeI_4	0.01	1.20	1.34	2.54	2.49	0.05
PI_3	0.03	1.06	1.35	2.41	2.47	0.06
SbI_3	0.03	1.35	1.35	2.70	–	–
HI	0.04	0.28	1.36	1.64	1.61	0.03
SnI_4	0.04	1.31	1.36	2.67	2.64	0.03
PbI_4	0.04	1.42	1.36	2.78	2.77	0.01
BiI_3	0.04	1.44	1.36	2.80	2.84	0.04
BI_3	0.06	0.71	1.37	2.08	2.03	0.05
AlI_3	0.12	1.05	1.41	2.46	2.44	0.02
PbI_2	0.14	1.37	1.43	2.80	2.79	0.01
SnI_2	0.15	1.30	1.44	2.74	2.73	0.01
TiI_4	0.17	1.10	1.45	2.55	2.55	0.00
BeI_2	0.19	0.68	1.46	2.14	2.12	0.02
MgI_2	0.25	1.06	1.51	2.57	–	–
TlI	0.28	1.28	1.53	2.81	2.81	0.00
CaI_2	0.30	1.33	1.54	2.87	–	–
SrI_2	0.33	1.48	1.56	3.04	3.03	0.01
BaI_2	0.39	1.58	1.60	3.18	3.20	0.02
LiI	0.53	0.70	1.70	2.40	2.39	0.01
NaI	0.54	1.02	1.71	2.73	2.71	0.02
KI	0.63	1.27	1.77	3.04	3.05	0.01
RbI	0.65	1.39	1.79	3.18	3.18	0.00
CsI	0.69	1.47	1.82	3.29	3.32	0.03

TABLE 5-8

**Bond Lengths in Solid Fluorides from Radii of
Charged Atoms (AU)**

Compound	$-\delta_F$	r_M	r_F	$r_M + r_F$	R_{exp}	R_{diff}
CdF_2	0.27	1.39	0.93	2.32	2.32	0.00
BeF_2	0.29	0.72	0.95	1.67	1.61	0.06
AgF	0.38	1.42	1.03	2.45	—	—
MgF_2	0.41	1.09	1.06	2.15	2.02	0.13
CaF_2	0.47	1.22	1.11	2.33	2.36	0.03
SrF_2	0.50	1.48	1.14	2.62	2.51	0.11
BaF_2	0.57	1.58	1.21	2.79	2.68	0.11
LiF	0.74	0.74	1.36	2.10	2.01	0.09
NaF	0.75	0.97	1.37	2.34	2.31	0.03
KF	0.84	1.16	1.46	2.62	2.66	0.04
RbF	0.86	1.27	1.48	2.75	2.82	0.07
CsF	0.90	1.48	1.51	2.99	3.01	0.02

TABLE 5-9

**Bond Lengths in Solid Chlorides from Radii of
Charged Atoms (AU)**

Compound	$-\delta_{Cl}$	r_M	r_{Cl}	$r_M + r_{Cl}$	R_{exp}	R_{diff}
$CdCl_2$	0.21	1.40	1.24	2.64	2.65	0.01
$BeCl_2$	0.28	0.76	1.26	2.02	2.02	0.00
$CuCl$	0.29	0.93	1.34	2.27	2.35	0.08
$AgCl$	0.30	1.44	1.35	2.79	2.77	0.02
$MgCl_2$	0.34	1.14	1.39	2.53	2.54	0.01
$CaCl_2$	0.40	1.30	1.47	2.77	2.74	0.03
$SrCl_2$	0.43	1.54	1.50	3.04	3.02	0.02
$BaCl_2$	0.49	1.64	1.57	3.21	3.18	0.03
$LiCl$	0.65	0.81	1.76	2.57	2.57	0.00
$NaCl$	0.67	1.03	1.79	2.82	2.81	0.01
KCl	0.76	1.23	1.90	3.13	3.14	0.01
$RbCl$	0.78	1.35	1.92	3.27	3.29	0.02
$CsCl$	0.81	1.57	1.95	3.52	3.50	0.02

TABLE 5-10

Bond Lengths in Solid Bromides from Radii of Charged Atoms (AU)

Compound	$-\delta_{Br}$	r_M	r_{Br}	$r_M + r_{Br}$	R_{exp}	R_{diff}
$CdBr_2$	0.17	1.42	1.35	2.77	2.76	0.01
$BeBr_2$	0.20	0.78	1.39	2.17	2.16	0.01
$CuBr$	0.25	0.99	1.45	2.44	2.46	0.02
$AgBr$	0.25	1.45	1.45	2.90	2.89	0.01
$MgBr_2$	0.31	1.16	1.53	2.69	2.70	0.01
$CaBr_2$	0.36	1.34	1.59	2.93	2.94	0.01
$SrBr_2$	0.39	1.58	1.62	3.20	3.21	0.01
$BaBr_2$	0.45	1.67	1.70	3.37	3.38	0.01
$LiBr$	0.61	0.84	1.90	2.74	2.75	0.01
$NaBr$	0.62	1.07	1.91	2.98	2.98	0.00
KBr	0.71	1.28	2.02	3.30	3.29	0.01
$RbBr$	0.73	1.40	2.05	3.45	3.43	0.02
$CsBr$	0.77	1.61	2.10	3.71	3.72	0.01

TABLE 5-11

Bond Lengths in Solid Iodides from Radii of Charged Atoms

Compound	$-\delta_I$	r_M	r_I	$r_M + r_I$	R_{exp}	R_{diff}
CdI_2	0.12	1.43	1.50	2.93	2.98	0.05
BeI_2	0.14	0.82	1.52	2.34	2.37	0.03
CuI	0.17	1.11	1.57	2.68	2.62	0.06
AgI	0.17	1.46	1.57	3.03	3.05	0.02
MgI_2	0.24	1.21	1.66	2.87	2.94	0.07
CaI_2	0.30	1.41	1.75	3.16	3.04	0.12
SrI_2	0.33	1.63	1.79	3.42	3.42	0.00
BaI_2	0.39	1.71	1.87	3.58	3.59	0.01
LiI	0.53	0.91	2.06	2.97	3.03	0.06
NaI	0.54	1.13	2.08	3.21	3.23	0.02
KI	0.63	1.36	2.20	3.56	3.53	0.03
RbI	0.65	1.48	2.23	3.71	3.66	0.05
CsI	0.69	1.69	2.28	3.97	3.96	0.01

TABLE 5-12

Bond Lengths in Gaseous Halides of 18-Shell Elements (AU)

Compound	$-\delta_X$	r_M	r_X	$r_M + r_X$	R_{exp}	R_{diff}
CuF	0.38	0.91	0.88	1.79	1.74	0.05
CuCl	0.29	0.91	1.20	2.11	–	–
CuBr	0.25	0.91	1.31	2.22	–	–
CuI	0.17	0.91	1.45	2.36	2.40	0.04
AgF	0.38	1.07	0.88	1.95	–	–
AgCl	0.30	1.07	1.21	2.28	2.25	0.03
AgBr	0.25	1.07	1.31	2.38	–	–
AgI	0.17	1.07	1.45	2.52	2.54	0.02
ZnF_2	0.23	1.01	0.80	1.81	1.81	0.00
$ZnCl_2$	0.16	1.01	1.11	2.12	2.09	0.03
$ZnBr_2$	0.13	1.01	1.23	2.24	2.24	0.00
ZnI_2	0.07	1.01	1.38	2.39	2.42	0.03
CdF_2	0.27	1.13	0.82	1.95	–	–
$CdCl_2$	0.21	1.13	1.14	2.27	2.24	0.03
$CdBr_2$	0.17	1.13	1.26	2.39	2.39	0.00
CdI_2	0.12	1.13	1.41	2.54	2.56	0.02
HgF_2	0.23	1.18	0.80	1.98	–	–
$HgCl_2$	0.17	1.18	1.11	2.29	2.29	0.00
$HgBr_2$	0.14	1.18	1.24	2.42	2.41	0.01
HgI_2	0.08	1.18	1.39	2.57	2.59	0.02
GaF_3	0.15	1.08	0.76	1.84	1.88	0.04
$GaCl_3$	0.10	1.08	1.06	2.14	2.09	0.05
$GaBr_3$	0.08	1.08	1.20	2.28	–	–
GaI_3	0.04	1.08	1.36	2.44	2.44	0.00
InF_3	0.19	1.38	0.78	2.16	–	–
$InCl_3$	0.14	1.38	1.09	2.47	2.46	0.01
$InBr_3$	0.11	1.38	1.22	2.60	2.58	0.02
InI_3	0.07	1.38	1.38	2.76	2.80	0.04

double repulsion effects prevented his ionic radii from being strictly additive in these crystals. By considering these effects carefully, however, he was able to account for the internuclear distances with extraordinary accuracy. But this work assumed the ionic model for such solids. According to the coordinated polymeric concept proposed herein, in which partially charged atoms instead of ions make up the crystal, the radius ratio is changed so that in no case is it 0.414 or smaller, which theoretically would allow anion–anion contact. The following are the smallest values calculated: LiI, 0.442; LiF, 0.544; and NaI, 0.553. The largest differences between calculated and experimental bond lengths are observed in fluorides and iodides. These differences may result from radius ratio differences but apparently not in the same way as assumed for ions.

The average difference between Pauling's ionic radius sums and the observed bond lengths in the twenty alkali metal halides is 0.07 Å, whereas computed on the basis of the partially charged atoms, it is only 0.03 Å. It is especially noteworthy that structural differences in solids do not appear to exert very significant effects on the bond lengths, for the calculated values make no structural distinction. Included in the tabulated data are compounds exhibiting a variety of environmental conditions for the atoms.

The extensive polarization of negative hydrogen, as discussed later, prevents accurate determinations of bond length in its binary compounds by calculation. However, when the charge on hydrogen is only slightly negative, or positive, the radii and bond lengths can be successfully treated, as shown in Table 5-13.

Bond lengths in solid chalcides can be determined somewhat less accurately, as shown in Table 5-14.

Other Compounds

All the binary compounds considered up to this point consist of molecules or nonmolecular solids in which all bonds are expected to be exactly equivalent. These appear to be the only bonds for which a pseudo-sphericity of individual atoms can

TABLE 5-13

Bond Lengths in Binary Hydrogen Compounds from Radii of Charged Atoms (AU)

Compound	$-\delta_H$	r_H	r_E	$r_H + r_E$	R_{exp}	R_{diff}
HF	0.25	0.08	0.81	0.89	0.92	0.03
HCl	0.16	0.16	1.11	1.27	1.27	0.00
H_2O	0.12	0.20	0.76	0.96	0.96	0.00
HBr	0.12	0.20	1.22	1.42	1.41	0.01
H_2S	0.05	0.27	1.06	1.33	1.33	0.00
H_2Se	0.05	0.27	1.20	1.47	1.46	0.01
H_3N	0.05	0.27	0.75	1.02	1.02	0.00
HI	0.04	0.28	1.36	1.64	1.61	0.03
H_3As	0.02	0.30	1.22	1.52	1.52	0.00
H_4C	0.01	0.31	0.79	1.10	1.09	0.01
GeH_4	0.00	0.32	1.22	1.54	1.54	0.00
PH_3	−0.01	0.33	1.09	1.42	1.42	0.00
SbH_3	−0.01	0.33	1.37	1.70	1.71	0.01
SnH_4	−0.02	0.34	1.35	1.69	1.70	0.01
SiH_4	−0.04	0.36	1.08	1.44	1.46	0.02

TABLE 5-14

**Bond Lengths in Solid Chalcides from Radii of
Charged Atoms (AU)**

Compound	$-\delta_X$	r_M	r_X	$r_M + r_X$	R_{exp}	R_{diff}
BeO	0.36	0.79	0.84	1.63	1.60	0.03
MgO	0.50	1.21	0.90	2.11	2.10	0.01
CaO	0.56	1.43	0.92	2.35	2.40	0.05
SrO	0.60	1.65	0.94	2.59	2.57	0.02
BaO	0.68	1.74	0.97	2.71	2.76	0.05
BeS	0.24	0.83	1.20	2.03	2.10	0.07
MgS	0.38	1.25	1.29	2.54	2.54	0.00
CaS	0.44	1.50	1.33	2.83	2.83	0.00
SrS	0.48	1.70	1.36	3.06	3.00	0.06
BaS	0.52	1.80	1.38	3.18	3.18	0.00
BeSe	0.24	0.78	1.32	2.10	–	–
MgSe	0.38	1.25	1.41	2.66	2.72	0.06
CaSe	0.46	1.49	1.47	2.96	2.96	0.00
SrSe	0.49	1.70	1.49	3.19	3.11	0.08
BaSe	0.53	1.80	1.51	3.31	3.31	0.00
BeTe	0.17	0.82	1.47	2.29	–	–
MgTe	0.31	1.18	1.56	2.74	–	–
CaTe	0.38	1.53	1.61	3.14	3.17	0.03
SrTe	0.42	1.73	1.64	3.37	3.33	0.04
BaTe	0.46	1.82	1.67	3.49	3.50	0.01

usefully be assumed. When the bonds within a compound are not all alike, then correct bond lengths are not obtained by summing the calculated radii of the charged atoms.

This becomes clearly evident in molecules such as those of methanol, wherein the observed H_3C-OH bond length is 1.43 Å. The equalization of electronegativity would produce a charge of -0.29 on oxygen but only 0.01 on carbon, since the electrons for the oxygen are in effect supplied by the hydrogen atoms, which acquire a charge of 0.07 each. According to this information, the contraction of the carbon radius from its nonpolar value of 0.77 Å would be negligible, but the oxygen radius should increase from the nonpolar 0.70 to 0.77 Å. The sum, 1.54 Å, is 0.11 Å longer than is experimentally observed. However, the average of the two partial charges is 0.15. If the radii of both carbon and oxygen are calculated as though their charges were 0.15 and -0.15, the values of 0.70 and 0.74 are obtained. Their sum, 1.44 Å, is in excellent agreement with the experimental bond length of 1.43 Å.

That this is not a fortuitous and isolated example is evidenced by the data of Table 5-15, which lists 26 examples including alcohols, ethers, thiols, thioethers,

TABLE 5-15

Bond Lengths in Alkyl Compounds from Charges and from
Ionic Blending Coefficients

Compound and bond	δ_A	δ_B	t_i	$r_A(\delta)$ $r_A(t_i)$	$r_B(\delta)$ $r_B(t_i)$	Bond length Calc.	Bond length Exp.
CH_3-OH	0.01	−0.29	0.15	0.77	0.77	1.54	−
				0.70	0.74	1.44	1.43
C_2H_3-OH	−0.01	−0.31	0.15	0.77	0.77	1.54	−
				0.70	0.74	1.44	1.43
C_3H_7-OH	−0.02	−0.31	0.15	0.78	0.77	1.55	−
				0.70	0.74	1.44	1.45
C_4H_9-OH	−0.02	−0.32	0.15	0.78	0.78	1.56	−
				0.70	0.74	1.44	1.44
$(CH_2-OH)_2$	0.02	−0.28	0.15	0.77	0.77	1.54	−
				0.70	0.74	1.44	1.43
$(CH_3-)_2O$	−0.01	−0.31	0.15	0.77	0.77	1.54	−
				0.70	0.74	1.44	1.42
$C_2H_5-OCH_3$	−0.02	−0.31	0.15	0.78	0.77	1.55	−
				0.70	0.74	1.44	1.45
$(C_2H_5-)_2O$	−0.02	−0.32	0.15	0.78	0.78	1.56	−
				0.70	0.74	1.44	1.43
CH_3-SH	−0.03	−0.10	0.07	0.78	1.06	1.84	−
				0.75	1.05	1.80	1.82
C_2H_5-SH	−0.03	−0.11	0.07	0.78	1.06	1.84	−
				0.75	1.05	1.80	1.81
$(CH_3-)_2S$	−0.03	−0.11	0.07	0.78	1.06	1.84	−
				0.75	1.05	1.80	1.82
$(C_2H_5-)_2S$	−0.03	−0.11	0.07	0.78	1.06	1.84	−
				0.75	1.05	1.80	1.81
$CH_3\ NH_2$	−0.02	−0.18	0.08	0.78	0.75	1.53	−
				0.73	0.74	1.47	1.47
$(CH_3-)_2NH$	−0.03	−0.18	0.08	0.78	0.75	1.53	−
				0.73	0.74	1.47	1.46
$(CH_3-)_3N$	−0.03	−0.19	0.08	0.78	0.75	1.53	−
				0.73	0.74	1.47	1.46
$C_2H_5-NH_2$	−0.03	−0.18	0.08	0.78	0.75	1.53	−
				0.73	0.75	1.48	1.46
$(C_2H_5-)_2NH$	−0.03	−0.19	0.08	0.78	0.75	1.53	−
				0.73	0.74	1.47	1.47
$(C_2H_5-)_3N$	−0.04	−0.19	0.08	0.79	0.75	1.54	−
				0.73	0.74	1.47	1.47
CH_3-Cl	0.08	−0.18	0.13	0.73	1.12	1.85	−
				0.71	1.08	1.79	1.78
C_2H_5-Cl	−0.01	−0.25	0.12	0.77	1.17	1.94	−
				0.71	1.08	1.79	1.78
C_3H_7-Cl	−0.02	−0.26	0.12	0.78	1.18	1.96	−
				0.71	1.08	1.79	1.75

TABLE 5-15 (continued)

Compound and bond	δ_A	δ_B	t_i	$r_A(\delta)$ $r_A(t_i)$	$r_B(\delta)$ $r_B(t_i)$	Bond length Calc.	Bond length Exp.
CH_3-Br	0.00	−0.17	0.08	0.77	1.26	2.03	−
				0.73	1.20	1.93	1.94
C_2H_5-Br	−0.02	−0.17	0.08	0.78	1.26	2.04	−
				0.73	1.20	1.93	1.94
C_3H_7-Br	−0.02	−0.19	0.09	0.78	1.27	2.05	−
				0.73	1.20	1.93	1.91
CH_3-I	−0.03	−0.05	0.01	0.78	1.37	2.15	−
				0.77	1.34	2.11	2.14
C_2H_5-I	−0.04	−0.05	0.01	0.78	1.37	2.15	−
				0.77	1.34	2.11	2.18

amines, and halides. These data suggest that sphericity may not be retained under conditions in which the bonds to a given atom are not alike.

In more complex situations, simple summing of calculated atomic radii fails to produce accurate bond lengths. In view of the evidence cited earlier in this chapter, it seems probable that the partial charges on the combined atoms may well be one factor in determining bond length, but that in more complex compounds, other factors are also involved, which as yet have not been fully identified or evaluated.

SIX

Polar Covalence II:
The Calculation of Polar Bond Energy

THE NEED FOR BOND ENERGY CALCULATION

The Significance of Bond Energy

All of the physical properties of matter depend on the kinds of atoms it contains and how they are arranged. All of the chemical properties of matter depend on the possibilities for atomic rearrangement. At ordinary temperatures perhaps the most important factor influencing the atomic rearrangement, which we call chemical change, is bond energy. Except for monatomic gases, all substances consist of atoms joined together primarily by chemical bonds. Atomic rearrangements tend to occur in the direction of formation of stronger bonds. An understanding of chemical bonds must include a quantitative accounting of their strength through an accurate recognition of the factors which produce that strength. It is not enough to be able to evaluate bond strength experimentally. We must also be able to predict it, or to represent the true experimental value through a reasonably simple physical or conceptual model. Bond energies are vitally important in determining the possibilities and, to a considerable extent, the mechanisms of chemical reactions. Nothing is closer to the heart of chemistry than bond energy.

The Problem of Calculating Bond Energy

Each individual atom consists of a structured arrangement of electrons and nucleus which holds together through a complex system of attractions and repulsions that results in a substantially lower energy content than that of the component particles if they existed independently. To obtain by calculation the atomization energy or the bond energy of a compound, one would need to be able to know or calculate the total electronic energy of the compound and also of the individual atoms. The difference would be the total bond energy.

The nature of the problem can be better understood by example. We can obtain the total electronic energy of an atom by summing the separate ionization energies necessary to strip the atom down to its bare nucleus. These energies are available from experiment only for the first twenty elements. However, it is possible to conduct extensive extrapolations which lead to reasonable estimates of ionization energies not yet available from experiment. In this manner, I have succeeded in estimating all the missing ionization energies for all the chemical elements through xenon, atomic number 54. The first four or five successive ionizations of a given atom have energies too complex to be estimated reliably, but these are usually available from experiment. Thereafter, successive values follow a much simpler pattern. I have found that, from boron on, the total electronic energy of an atom is a simple function of its atomic number:

$$E = AZ^B$$

As determined from a least squares treatment, $A = 13.8418$ and $B = 2.40276$. This equation is then equivalent to the linear log–log relationship,

$$\log E = 2.40276 \log Z + 1.14120$$

For the elements from boron through calcium the sums of the experimental ionization energies, and from calcium through xenon the sums of the experimental plus estimated ionization energies, agree with the total electronic energies calculated by the above equation to within 0.25% on the average, the maximum difference being 0.6%.

According to this equation, the total electronic energy of a potassium atom is 377,225 kpm, and of a bromine atom, 1,622,919 kpm. The bond energy of a potassium bromide gas molecule is about 91 kpm. The total electronic energy of a KBr molecule* is the sum of that for potassium and for bromine plus 91 kpm, or 2,000,235 kpm. This is a typical situation in that the energy of bonding is almost invariably far smaller than the total energies of the separate or combined atoms. Consequently, the bond energy would have to be determined as the difference, a small difference, between two relatively very large numbers. In KBr(g), the molecule is 0.0045% more stable than the separate atoms! To calculate the bond energy as the difference between the total molecular and atomic energies within, let us say,

---

*Including internuclear interactions.

even *10%* would require each separately to be accurate within 5%—and that means 5% of the bond energy, not of the total energy. 5% of 91 is 4.55 kpm, which is 0.0002% of two million. And, of course, a bond energy accurate within *1%* would require calculation of the total energy within 0.00002%. On the other hand, quantum mechanical accuracy within 1% would yield the bond energy within the limits of plus or minus about three orders of magnitude.

This discussion should suffice to persuade us that, especially since exact calculations on systems of more than two particles are presently impossible, there is little hope of finding any quantum mechanical method of calculating bond energy accurately. Even if it could be done, one would hardly expect the method to be simple and practical and able to contribute much attainable understanding of chemical bonds. The temptation to keep trying is very understandable, since if we could begin with the component particles and simply calculate exactly their interactions, we would then be able to arrive at our goal without the necessity of invoking such concepts as electronegativity or covalent radius or effective nuclear charge. But at least for the foreseeable future, any progress toward understanding chemical bonds seems unquestionably dependent on the successful invocation of simpler concepts and on the intuitive recognition of various relationships.

A recent and unusually lucid statement of the present condition of quantum mechanical approaches to these problems is presented by Dewar.[1]

THE CALCULATION OF HETERONUCLEAR BOND ENERGY

The Model of Polar Covalence

The unreactiveness of neutral atomic structures in which an outermost shell of two or eight electrons is featured demonstrates the effectiveness with which the nucleus is shielded in these structures. We are therefore somewhat justified in considering primarily the outermost principal quantum level in bonding, in a sense disregarding the underlying electrons. To this extent, the concept of two otherwise quite independent atoms being held together by mutual sharing of a pair of outermost electrons appears reasonable. Even though molecular orbital theory would imply that all the electrons are mutually held by the two nuclei, assuming a diatomic molecule under consideration, the approximation of two atoms tangentially in contact is acceptable if only two bonding electrons are involved in the bond.

Thus it is with the full realization that the concept is crude and oversimplified that we accept the simpler picture and make what use we can of the separate

[1] M. J. S. Dewar, *Science* **187,** 1037 (1975); see also **190,** 591 (1975).

behavior of each atom when a polar covalent bond is formed. We shall speak of individual atoms within a chemical compound despite the assertions of certain theorists that atoms do not exist as such under conditions of chemical combination. We shall imagine them as possessing distinguishably separate electronic populations even though the electrons are alleged to belong to the molecular or nonmolecular compound system as a whole. We shall picture them as contracting and expanding according to their individual electronic populations and as acquiring partial charges both positive and negative. We shall calculate the bond energy as though the compound were a blend of two hypothetical extremes, neither of which is ever realized, even momentarily, in a significant manner unless required by some outside agent.

In other words, we define that part of a molecular system which contains two atoms bonded together, not as the complex assemblage which it actually must be, but as a measurable blend of the two imaginary extremes. The physical picture of the polar covalent bond has already been painted. We have seen that, according to this picture, the bonding electrons are induced to spend more than half the time more closely associated with that nucleus which originally attracted them more strongly and less than half the time more closely associated with that nucleus which originally attracted them less. Despite the fact that both electrons belong to both atoms, we apportion them, in effect, by saying that the population of one atom consists of its original cloud plus a measured fraction of an electron from the other atom and that the population of the other atom consists of its original cloud minus that fraction lost to the first atom. Thus the first atom now bears a partial negative charge, leaving the second atom partially positive. Let us suppose that the fraction of charge transferred is 0.25 electron. Then the electron population on the initially more electronegative atom is $Z_1 + 0.25$, and the electron population on the initially less electronegative atom is $Z_2 - 0.25$, where Z_1 and Z_2 are the atomic numbers of the two elements.

Instead of calculating the ionic energy resulting from the constant electrostatic attraction between negative and positive partial charges, we find it more appropriate in most applications to consider the bond as though it were completely nonpolar 75% of the time but completely ionic 25% of the time. Thus we are spared the complexity of dealing simultaneously with the two energy contributions and can handle them simply one at a time. Our model of polar covalence thus becomes essentially a dynamic model, actually intermediate in nature between nonpolar and ionic but, for simplicity in calculation, treated as though oscillating between the two extremes but with zero oscillating time.

The simplicity of practice of applying this model is fantastic and almost unbelievable compared to the more fashionable approaches of quantum mechanics, but it is amply justified by the success of the results. There may be some who will be tempted to ascribe this success to a remarkable mutual cancellation of errors. My own experience suggests that the law of averages is grossly misleading when applied to mathematical results, in which errors, rather than cancel one another half the time, are almost invariably additive!

Nonpolar Covalent Energy

In Pauling's original work on the evaluation of electronegativity, he assumed that a heteronuclear bond, if nonpolar, would have an energy exactly the average of the two homonuclear bonds. He found the arithmetic mean a more expedient average but suggested some preference for the geometric mean. When two values being averaged are of similar magnitude, there is no very significant difference between arithmetic and geometric means. When they are substantially different, however, then the geometric mean favors the smaller value. I cannot at this time fully justify the use of the geometric mean except on the pragmatic grounds that it appears to work very satisfactorily.

Pauling did not, however, make any correction of nonpolar bond energy for bond length. Any actual polar covalent bond tends to be appreciably shorter than the sum of the nonpolar covalent radii of the two atoms. This should enhance the nonpolar bond energy contribution, since shorter bonds are always stronger. If the bond should be longer than the covalent radius sum, this would similarly lessen the covalent bond energy contribution. I have therefore preferred to include a correction factor, which is merely the ratio of the nonpolar covalent radius sum to the observed bond length, R_c/R_0.

In summary, the nonpolar covalent energy of a heteronuclear bond is obtained, following Pauling but adding a correction factor for bond length, by the equation

$$E_c = \frac{R_c (E_{AA} E_{BB})^{1/2}}{R_0}$$

For assistance in determining geometric mean bond energies for various elemental combinations, the data of Table 6-1 are provided. They cover some of the more commonly occurring bonds.

The above is generally applicable for all bonds in molecular substances. For nonmolecular solids, where we wish to calculate the total atomization energy as opposed to the attraction between two neighbors, we must recognize that the electrons used in the bonding may not be limited to those useful in molecular bonding but may include all the external electrons of the atoms. This is because, in nonmolecular solids, the coordination is much more complete, normally univalent or divalent atoms becoming directly attached to several other atoms at the same time. We therefore include the factor n, which represents the equivalent number of bonding pairs involved per formula unit of the compound. The covalent energy contribution becomes the same as above, multiplied by n (usually of value 4). A fuller discussion of bonding in nonmolecular solids is to be found later in this chapter.

Ionic Energy

For calculating the bond energy in molecular substances, the value of the ionic energy E_i is simply the electrostatic energy between opposite unit charges at

TABLE 6-1

Parameters for Calculation of Covalent Energy $R_c (E_{AA}E_{BB})^{1/2}$

	H^a	F'	Cl'	Br'	I'	O'	S'
Li	90.1	67.5	94.6	89.5	85.4	62.8	94.0
Na	80.5	59.3	82.0	77.2	73.2	55.1	81.3
K	84.6	60.2	81.7	76.6	71.7	55.9	80.7
Rb	89.0	62.8	84.7	78.9	74.0	58.3	83.5
Cs	89.2	62.4	83.5	77.5	72.1	57.6	82.4
Be	90.2	71.7	104.1	100.0	96.8	67.0	104.5
Mg	102.0	76.2	106.4	100.5	95.7	70.7	105.5
Ca	123.0	89.1	121.8	114.3	107.8	82.7	120.7
Sr	130.0	93.0	126.4	118.3	111.1	86.1	124.8
Ba	117.1	84.3	113.2	105.8	99.3	77.5	111.7
Zn	108.9	82.0	115.2	109.1	104.1	76.2	114.4
Cd	103.6	76.6	106.6	100.6	95.4	71.3	105.8
Hg	–	39.9	55.6	52.3	49.6	37.2	54.9
B	96.2	78.0	114.2	110.2	106.9	72.8	112.5
Al	108.2	81.9	115.2	109.4	104.4	76.0	114.3
Ga	109.8	83.0	116.8	110.9	105.9	77.2	116.2
In	107.5	78.5	109.1	103.3	98.3	72.8	108.4
Tl	–	57.2	79.5	75.2	71.4	53.2	78.9
C	101.0	83.1	122.5	118.2	115.1	77.6	122.4
Si	111.2	84.9	120.3	114.1	109.8	79.1	119.8
Ge	110.1	83.6	118.2	112.3	107.4	78.0	117.5
Sn	107.2	79.7	111.1	105.2	99.9	74.1	110.3
Pb	82.7	61.3	84.9	80.1	76.2	56.9	84.3
N	66.0	55.8	82.7	79.9	77.8	52.1	82.6
P	103.7	79.9	113.9	108.6	104.5	74.5	113.4
As	100.6	77.4	109.7	108.6	99.8	72.0	109.0
Sb	98.1	73.3	102.4	96.8	92.1	68.0	101.6
Bi	100.2	74.3	108.3	97.2	92.3	68.9	102.3
O	56.5	50.2	74.7	72.3	70.6	–	74.6
S	100.4	80.2	114.9	109.9	105.3	74.6	–
Se	97.7	76.7	108.8	103.5	99.1	71.4	108.2
Te	99.4	74.3	104.1	98.6	93.8	66.0	103.2
F	55.6	–	80.0	77.7	39.8	50.2	80.2
Cl	93.5	80.0	–	110.3	106.3	74.7	114.9
Br	94.9	77.7	110.3	–	100.8	72.3	109.9
I	99.2	39.8	106.3	100.8	–	70.6	105.3

^{a}Use only for familiar binary molecules; other H compounds require individual corrections E_{HH}.

the distance of the bond length:

$$E_i = \frac{332}{R_0}$$

The factor 332 simply converts to units of kpm, the charge product being unity and the bond length being expressed in angstrom units.

In nonmolecular substances, the ionic energy is calculated using the Born–

Mayer equation for U, the crystal energy:

$$U/f = \frac{332Mkz + z - e^2}{R_0}$$

In this equation, M is the Madelung constant for the crystal, and k is the repulsion coefficient. The factor f is 1.00 for univalent "anions" but 0.63 for oxides, 0.57 for sulfides, and 0.54 for selenides, as will be discussed later.

Total Energy

According to the simple physical model of polar covalence described above, the total energy is represented simply as the weighted, or blended, sum of the two extreme forms of energy, covalent and ionic. Where the blending coefficients are t_c for the covalent contribution and t_i for the ionic contribution, then $t_c + t_i = 1.00$. The blending coefficients can be evaluated directly from the partial charges, for molecular compounds, according to the relationship

$$t_i = \frac{\delta_A - \delta_B}{2}$$

For nonmolecular solids then, t_i is the true percentage of ionicity as determined by partial charge.

In multiple bonding the entire energy is multiplied by the factor m, which has the value 0.50 for half-bonds, 1.00 for single bonds, 1.50 for double bonds, and 1.75 for triple bonds.

The total bond energy, or the atomization energy of a nonmolecular solid, can then be expressed as

$$E = t_c E_c n + t_i U/f$$

For a bond in a molecule, the total energy is

$$E = m(t_c E_c + t_i E_i)$$

Several points are worth emphasizing here.

First, as will be illustrated a little later, the method is extraordinarily successful in dealing quantitatively with a large number of chemical bonds. Thus it provides an insight not previously available, it permits a greater depth of understanding than has been possible before, and it even allows prediction and understanding of many chemical phenomena not previously well understood.

Second, it allows us a quantitative understanding of the fundamental importance of electronegativity and the differences among different elements.

For convenience in calculations, Table 6-1 provides the products of nonpolar covalent radius sums and geometric mean homonuclear single covalent bond energies for a number of common combinations of elements. Comparison of the magnitude of those products with the factor 332 shows us why ionicity of a bond always enhances its strength. Notice that the total energy of the bond, multiplied by the bond length experimentally observed, is equal to the blended or weighted sum of the just-mentioned product and the factor 332. There is no actual geometric mean homonuclear single covalent bond energy that, when multiplied by the nonpolar covalent radius sum, can equal 332. Consequently the ionic fraction of the energy always exceeds the covalent fraction which it displaces, with respect to quantity of energy. Since this is such a very important point, it is further emphasized by Fig. 6-1, in which an average or typical bond energy, purely covalent, of 60 kpm is shown to change with increased ionicity of the bond in an upward direction.

For this reason most displacement reactions, for example, involve breaking bonds to form more polar bonds, which means stronger bonds. For this reason compounds appear preferred over pure elements in the natural state, for compounds are nearly always more stable because of the bond polarity absent in the pure elements. A knowledge of initial electronegativities therefore equips one to make valid predictions covering a wide range of chemistry.

Third, the expression for total bond energy allows a means of predicting the blending coefficients in an entirely independent manner without any recourse to

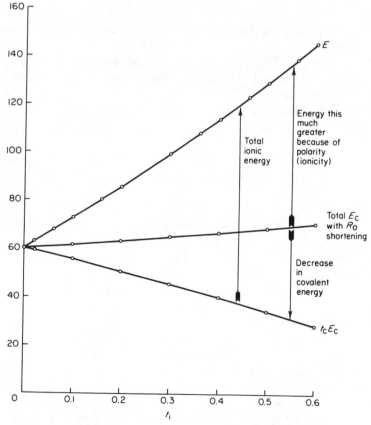

Fig. 6.1 Effect of ionicity of bond energy. Assumed: geometrical mean E is 60 kpm; R_c is 2.00 A; R_0 at 60% ionicity is 1.70 A.

electronegativity or its equalization or to partial charge. All we need do is accept the validity of the separate equations for the covalent and ionic energies. Then we have the equation

$$t_i = \frac{E - E_c}{E_i - E_c}$$

Calculation of partial charges from t_i is of course a simple matter. For example, in a compound of formula AX_3, the partial charge on X is one-third that on A. If the charge on X is a, then the charge on A is $3a$ (only of opposite sign). The average, $2a$, is the ionic blending coefficient t_i.

Table 6-2 gives, for 36 representative binary compounds both molecular and nonmolecular, the results of calculating partial charges both from bond energies and

TABLE 6-2

**Partial Charges from Experimental Bond
Energies (I) and from Electronegativities (II)
and Some Representative Calculated and
Experimental Atomization Energies**

Compound	Partial charges (I)	Partial charges (II)	Atomization energies Calc.	Atomization energies Exp.
CH_4 (g)	−0.05	−0.05	396.4	397.6
	0.01	0.01		
CS_2 (g)	0.04	0.05	278.8	276.4
	−0.02	−0.03		
ClF_3 (g)	0.13	0.13	121.1	124.7
	−0.04	−0.04		
NH_3 (g)	−0.16	−0.16	280.2	280.3
	0.05	0.05		
SO_2 (g)	0.17	0.16	254.3	256.7
	−0.09	−0.08		
HCl(g)	0.16	0.16	103.7	103.3
	−0.16	−0.16		
$ZnCl_2$ (c)	0.32	0.32	188.3	188.6
	−0.16	−0.16		
AgI(c)	0.15	0.17	110.2	108.3
	−0.15	−0.17		
BeO(c)	0.36	0.36	280.9	281.0
	−0.36	−0.36		
H_2O(g)	0.12	0.12	221.6	221.6
	−0.25	−0.25		
ZnF_2 (c)	0.46	0.46	244.4	244.8
	−0.23	−0.23		
$MnBr_2$ (c)	0.44	0.46	213.5	209.4
	−0.22	−0.23		
$PbCl_2$ (c)	0.46	0.46	191.9	191.2
	−0.23	−0.23		

TABLE 6-2 (continued)

Compound	Partial charges		Atomization energies	
	(I)	(II)	Calc.	Exp.
HF(g)	0.25	0.25	135.5	135.8
	−0.25	−0.25		
$SiCl_4$ (g)	0.45	0.43	380.0	382.3
	−0.11	−0.11		
B_2O_3 (g)	0.32	0.34	653.4	649.5
	−0.22	−0.23		
CaO(c)	0.60	0.56	245.9	253.9
	−0.60	−0.56		
PbF_2 (c)	0.58	0.58	243.0	244.2
	−0.29	−0.29		
BeF_2 (c)	0.58	0.58	349.7	359.4
	−0.29	−0.29		
SrI_2 (c)	0.68	0.66	221.8	225.5
	−0.34	−0.33		
BF_3 (g)	0.53	0.53	469.3	463.0
	−0.18	−0.18		
$CaBr_2$ (c)	0.72	0.72	258.2	257.3
	−0.36	−0.36		
TlBr(c)	0.36	0.36	111.4	111.5
	−0.36	−0.36		
CaF_2 (c)	0.94	0.94	369.9	370.6
	−0.47	−0.47		
AlF_3 (g)	0.72	0.72	428.4	421.2
	−0.24	−0.24		
$BaCl_2$ (c)	1.00	0.98	302.8	305.5
	−0.50	−0.49		
NaI(c)	0.56	0.54	118.4	120.2
	−0.56	−0.54		
KI(g)	0.64	0.63	77.3	77.8
	−0.64	−0.63		
RbI(c)	0.66	0.65	123.0	123.6
	−0.66	−0.65		
NaCl(c)	0.69	0.67	151.7	153.2
	−0.69	−0.67		
RbBr(g)	0.74	0.73	89.4	90.4
	−0.74	−0.73		
NaF(c)	0.74	0.75	183.3	181.8
	−0.74	−0.75		
KCl(g)	0.76	0.76	101.9	101.6
	−0.76	−0.76		
CsBr(c)	0.80	0.77	139.2	139.7
	−0.80	−0.77		
KF(c)	0.85	0.84	176.0	175.8
	−0.85	−0.84		

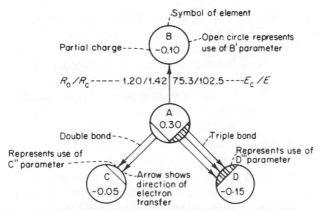

Fig. 6-2 Schematic representation of bonding in an imaginary molecule.

from electronegativities. The near-perfect agreement should come as no surprise in view of the successful calculation of bond energies, as also shown in Table 6-2. The data represent, however, the strongest supporting evidence of the validity of the partial charge values and the electronegativity scale upon which they are based. However reluctant one might be to accept partial charge values as atomic charges having any reasonable degree of reality, it must be difficult indeed to deny that partial charges do provide accurate blending coefficients for the calculation of bond energies. (See Fig. 6-2.)

Modifications of the Simple Method

Except for certain bonds which appear to be one-electron bonds, or "half-bonds," nearly all chemical bonds examined appear to have an energy that is either equal to or greater than the heteronuclear single covalent bond energy calculated as described in this chapter. If the bond energy is greater, this is an indication that either more than two electrons are shared in the bonding or that a reduction of the lone pair bond weakening effect described in Chapter Two has occurred. Multiplicity of bonding affects both covalent and ionic energies equally, as shown by the requirement of multiplying both energy contributions by the multiplicity factor. On the other hand, reduction of homonuclear single covalent bond weakening changes only the nonpolar covalent bond energy and thus has its greatest significance in bonds that are not highly polar. In highly polar bonds the factors influencing the covalent energy become less important.

NONMOLECULAR SOLIDS

Objections to the Ionic Model

Practically every textbook of chemistry published these many years says that there are two types of chemical bonding, covalent and ionic. In a covalent bond each atom supplies one electron and one vacancy to the other atom, and two electrons are shared by the two atoms. In an ionic bond each atom has the capacity to form covalent bonds, but one atom is much more electronegative than the other and therefore acquires both bonding electrons, leaving none with the other atom. Thus the first atom acquires unit negative charge and leaves the second atom with unit positive charge. The two ions attract one another by simple electrostatic attractions between opposite charges. Most books refer to "polar covalence" but not in a very informed manner. Most books hint that most actual chemical bonds are intermediate between covalent and ionic bonds but then go on using the extreme terms as though the intermediacy had been forgotten.

The temptation is great, especially when the ions produced by this remarkable transfer of a valence electron are isoelectronic with an M8 element, one of the helium family. It is nice to know why atoms react, and the existence of the M8 elements whose atoms in general remain strictly aloof suggests very easily that perhaps other elements tend to become like the M8 elements if they can. Thus the neon structure is attained when oxygen gains two electrons per atom, fluorine gains one electron per atom, sodium loses one electron per atom, magnesium loses two electron per atom, and so forth. Thus are accounted for Na_2O, MgO, NaF, MgF_2, etc. All the atoms are "trying to become like" neon.

Unfortunately, only in number of electrons do these ionic species in any way resemble neon. Otherwise, there is no similarity whatever, other than "nuclear model," among Na^+, Mg^{++}, Ne, F^-, and $O^=$. This seems not very bothersome to most chemists, whose concepts of chemistry may become gradually more sophisticated with training and experience but who commonly remain perfectly willing to think in terms of ions in the crystalline solids which they form.

Perhaps the most flagrant defect in the concept of ionic solids is its failure to recognize the most important fundamental properties of ions. A simple cation is a positive charge surrounded by an electronic cloud insufficient to balance the positive charge. It therefore contains outer vacant orbitals within which appreciable effective nuclear charge can be sensed by an electron or electron pair. Thus a simple cation is inherently a potential electron pair acceptor, prepared for coordination at any opportunity. A simple anion always has an outermost level of electron pairs, which mutually shield one another from the nucleus to the extent that the effective nuclear charge is not very large. The surplus of electrons makes it easy for this species to serve as potential electron pair donor in coordination. Furthermore, the anion, like the cation, is also a positive charge surrounded by a cloud of electrons, this time more than sufficient to balance the nuclear charge. Although it is convenient to imagine the charges on cations and anions to be centered at their nuclei and the attractions to be electrostatic attractions between the opposite charges at the distance of the internuclear separation, it is certainly not realistic. It would not be realistic unless the ions were separated much farther apart, whereas actually they appear to be in direct contact. Therefore the attraction is not between nuclear centers of opposite charge, but, just as in covalence, it is the attraction of both nuclei for the same electrons. The only difference is that, in the ionic model, the bonding electrons are claimed to "belong" to the anion exclusively.

Call it what we will, we cannot escape the fact that potential electron pair acceptors cannot possibly be brought in direct contact on all sides with potential electron pair donors, and vice versa, without coordination occurring. But in an "ionic" crystal, that is exactly how the atoms are stacked. I would have no objection whatever to the assertion that an "ionic" crystal would be formed by condensation of oppositely charged gaseous ions, provided it is understood that the term "ionic" refers to the origin and not the product. It seems much more logical to consider such a solid as involving coordination among the ions. I have therefore suggested the alternative name, **coordinated polymeric** model.

It is interesting to note that the sum of the "ionic radii" so commonly used to estimate the internuclear distance in such a crystal is, more frequently than not, very similar in quantity to the sum of the nonpolar covalent radii. Thus the bond length itself adds no strength whatever to the concept of the nature of the bonding in these nonmolecular solids. Perhaps the distinction I would make between the two models is not one that can be tested adequately by experiment, but I believe it allows a reasonably clear mental differentiation. If the solid were truly ionic, then the bonding electrons, although electrostatically attracted by the surrounding nuclei of cations, would never penetrate toward such nuclei but remain exclusively a part of the anion system. But if the solid were coordinated polymeric, then the bonding electrons, although largely monopolized by the nucleus of the anion, would also penetrate toward the cation nuclei to some extent, imparting a covalent character to the "ionic" bonding.

In the first edition of this book, it was mentioned that when a careful X-ray examination of a crystal is made, it is sometimes possible to calculate an electron density variation between cation nucleus and anion nucleus. This has a minimum value, which, although not zero, seems to have no other interpretation than that it represents the junction of the two atoms (ions). This type of analysis has been performed for only a relatively few compounds, but in each case the atomic radii found in this manner are larger than expected for cations and smaller than expected for anions, by comparison with the "standard" ionic radii usually tabulated. Thus qualitatively the X-ray data are in agreement with the concept of only partially charged atoms rather than ions, assuming that the ionic radii are valid for the ions. Since that time, it has become possible, as described in Chapter Two, to estimate the atomic radii of charged atoms such that, when added, they produce the correct internuclear distance. The estimated radii do not agree with the X-ray data either. For example, in NaCl crystal the internuclear distance is experimentally 2.81 Å. This is well accounted for by the estimated atomic radius sums of 1.03 for the sodium and 1.79 for the chlorine that correspond to the calculation of 67% ionicity. But the X-ray "radii" are Na 1.18 Å and Cl 1.64 Å, and the "ionic radii" are Na^+ 0.98 Å and Cl^- 1.81 Å. Obviously there is confusion here which needs clarification. We can only conclude that "ionic radii" do not support the ionic model.

The ease of ion formation in water solution is also commonly accepted as evidence of the ionic nature of nonmolecular solids. However, the process of dissolution of a salt may equally well be thought of as merely a change in the coordination sphere. For example, in sodium chloride, chloride ions change from an environment of sodium ions to an environment of water molecules, the true ionic condition being modified in both situations. Likewise, sodium ions change from an environment of chloride ions to an environment of water molecules. The coordination bonding between ion and water is not very different from that between opposite "ions" in the crystal.

The experimental observations of the electrolysis of melts of nonmolecular solids also tend to confirm the "ionic" model. However, there are two objections to this line of reasoning. First, the temperatures required to melt a salt are usually sufficiently high as to make comparison with ordinary temperatures unreasonable. Evidence of ions at high temperatures does not prove their existence at ordinary temperatures. Second, the production of an electrode product that appears to have come from discharge of ions does not necessarily mean that the substance must have been present as ions. Consider the ease with which hydrogen and oxygen are obtained from water, when only a small concentration of electrolyte is present.

All in all, the reasons for believing in the ionic model are persuasive, attractive, and almost convincing, but they are not rigorously sound and they lead to a less accurate concept of nonmolecular solids than seems permissible.

Advantages of and Questions about the Coordinated Polymeric Model

With only a minimum of essential modification, the coordinated polymeric model removes the old and misleading dichotomy of "covalent" and "ionic" bonding and fits in smoothly with the new model of polar covalence. This allows for the first time a reasonably consistent bonding concept that extends to most of chemistry. This is particularly advantageous in application to such substances as magnesium oxide, in which the ionic model seems totally unrealistic in view of the large ionization energies needed to produce the dipositive magnesium ion, the large energy needed to force the oxygen atom to acquire two electrons, and the practical impossibility that the cations could be surrounded on all sides by such reluctant anions, or the anions by such electron-hungry cations, without coordination occurring.

Some typical results of calculating the atomization energy of nonmolecular solids are tabulated in later chapters, as well as in Table 6-2, which includes 70 solids.

Two difficulties which were pointed out in the first edition have still not yet been removed. One is the apparent fact that the effective number of bonding electron pairs in certain binary solids, including those of the smaller metal atoms, lithium, sodium, beryllium, magnesium, aluminum, and zinc, in their appropriate groups, seems smaller than the expected 4 for the value of n in the covalent energy equation, namely, 3. Although this difference appears to be supported by the data, it is not yet explained.

The other difficulty seems probably related to the repulsion coefficient k, used only in calculating the ionic energy contribution in nonmolecular solids. The methods of estimating k appear to be based on the assumption of the existence of the atoms as ions, and if they are not truly in ionic condition, then the value of k may be in considerable error. In fact, for all the chalcides, in which a dinegative anion is postulated, a correction factor must be used. Possibly this correction may not be related to the repulsion coefficient at all, but there is no other obvious source of its need.

A New and Different Model of Certain Nonmolecular Solids

The coordinated polymeric model of nonmolecular solids, even when corrected as discussed above, appears inadequate to describe certain types of one-to-one solids of special interest as semiconductors. Table 6-3 records the atomization energy calculations for certain 1:1 nonmolecular solids, based on an alternative, "electrostatic" model. In zinc oxide, for example, the atomization energy is calculated as the sum of two single bonds between zinc and oxygen, of a normal polar nature, to which is added the purely electrostatic attractions between the zinc

TABLE 6-3

Atomization Energies of Some Solids
as Calculated from Electrostatic
Model

| Compound | Atomization energy (kpm) | |
	Calc.	Exp.
ZnO	174.8	173.8
ZnS	142.6	145.8
ZnSe	123.6	119.2
CdO	150.4	147.5
CdS	130.4	128.
CdTe	95.7	94.6
AlN	265.7	266.
AlP	199.2	197.6
GaN	200.8	205.6
GaAs	157.8	155.5
InP	148.7	146.2
InSb	119.4	115.2
ZnTe	93.0	104.6
CdSe	123.6	108.8
AlAs	203.7	178.1
GaP	151.6	167.0
GaSb	118.2	138.9
InN	192.5	163.2
InAs	156.1	131.5

and another two oxygen atoms, based only on their calculated partial charges. Whereas most of the solids in Table 6-3 appear adequately described, some of the results are much in error and the cause is as yet unknown.

Although to date the atomization energies of more complex salts and oxides have not been explored in detail, there is some evidence that the new model may be appropriate for such common substances as calcium carbonate and sodium sulfate.

SEVEN

Chemical Combinations
of Hydrogen

THE SPECIAL NATURE OF HYDROGEN

We are sometimes inclined to regard hydrogen atoms as just another kind of atom, entering into chemical combination just as other kinds of atoms do. In fact, however, we should be cautious immediately. In all active elements except hydrogen, the valence electrons are superimposed upon a core in which more stably held electrons to some degree protect the nucleus. For hydrogen alone the valence electron *is* the entire electronic cloud, with nothing else in the atom except the bare proton. It is wise to suspect, therefore, that although each chemical element is "unique," hydrogen is "more unique" than any other.

Covalent Radius

This suspicion is strengthened by a study of the bond lengths in hydrogen compounds wherein the polarity is too small to exert much influence on the atomic radii. In the H_2 molecule the bond length of 0.74 Å suggests a nonpolar covalent radius of 0.37 Å, but the value most frequently observed is more nearly 0.32 Å. Consequently it is useful to adopt 0.32 as the covalent radius of hydrogen even though this differs from what the hydrogen molecule reveals. As described in Chapter Five, it has been possible to calculate bond lengths for single covalent bonds to hydrogen in which the hydrogen becomes fairly highly positive; but for negatively charged hydrogen, it is possible to calculate bond lengths only when the charge is small.

Electronegativity

So important an element as hydrogen ought certainly to be correctly evaluated as to electronegativity. Unfortunately, however, it has up to now been possible only to assign a reasonable value and test it in use. Pauling quite arbitrarily assigned a value of 2.1 on his scale according to which boron has the value 2.0 and carbon 2.5. This value has been widely accepted ever since. However, there are two implications of this value which seem unacceptable. One is that the difference between carbon and hydrogen is then quite appreciable and should lead to fairly substantial negative charge on carbon in hydrocarbons, with correspondingly substantial positive charge on hydrogen. In general, hydrogen attached to carbon in hydrocarbons has not been considered conspicuously positive. Furthermore, higher negative charge on carbon should cause expansion of the electronic sphere of carbon atoms and thus increase the carbon–carbon single bond length. In fact, it appears to be unchanged in alkanes at 1.54 Å, the same as in a crystal of diamond where the partial charge on carbon must be zero. The second implication of the 2.1 electronegativity of hydrogen is that it is thus only slightly higher than that of boron, and the B–H bond should be only very slightly polar with the hydrogen inconspicuously negative. In fact, in the simpler and more reactive boron hydrides, the B–H bond appears very readily hydrolyzed and easily susceptible to oxidation—in fact, characteristically involving partially negative hydrogen.

On the basis of these facts, it appeared more acceptable to assign to hydrogen an electronegativity somewhat closer to carbon and therefore appreciably higher than the value for boron. In the early work on evaluating electronegativity by the relative compactness method, it appeared desirable to give hydrogen an electronegativity of 3.55 on the scale in which boron was 2.84 (since then revised to 2.93) and carbon 3.79. This value of 3.55 corresponds to about 2.3 on the Pauling scale—not a large change but believed significant. As it has turned out subsequently, and will be demonstrated in this chapter, the chosen value of 3.55 seems very satisfactory, providing quantitative data of great accuracy in most applications.

Bond Energy Corrections

Early calculations of bond energies in binary compounds containing partially positive hydrogen gave results that were too high in proportion to the magnitude of the positive charge. An effective empirical correction is applied by multiplying the homonuclear single covalent bond energy of hydrogen, 104.2, by the factor $1.00 - \delta_H$, when δ_H is positive. Table 7-1 includes such binary compounds of positive hydrogen together with the calculated and experimental atomization energies.

Atomization energies of binary compounds containing slightly negative hydrogen can be calculated without correction, but when the hydrogen is more than slightly negative, the results are all too high. The explanation is somewhat speculative, but it relates to the reasonable expectation that highly negative hydrogen should be extremely polarizable. It is proposed that the polarization is so great that

TABLE 7-1

Atomization of Binary Hydrogen Compounds

Compound	δ_H	R_0	Atomization energy, kpm	
			Calc.	Exp.
LiH(g)	−0.49	1.60	56.4[a]	59.8
LiH(c)	−0.49	2.04	132.9[a]	112.1
NaH(g)	−0.50	1.89	42.6[a]	48.1
NaH(c)	−0.50	2.44	99.2[a]	91.7
KH(g)	−0.60	2.24	37.8[a]	43.5
KH(c)	−0.60	2.85	89.1[a]	88.0
RbH(g)	−0.63	2.37	37.6[a]	38.7
RbH(c)	−0.63	2.98	88.7[a]	92.0
CsH(g)	−0.65	2.49	35.8[a]	41.9
CsH(c)	−0.65	3.19	84.0[a]	82
BeH(g)	−0.23	1.30	69.4	53
BH(g)	−0.08	1.23	77.6[a]	79.1
BH$_3$(g)	−0.04	(1.14)	253.5[a]	266.8
B$_2$H$_6$(g)	−0.04	1.19	572.0	573.1
		1.33 (4 half-bonds)		
AlH(g)	−0.19	1.65	65.7[a]	68.1
GaH(g)	−0.03	1.58	69.6[a]	65.6
InH(g)	−0.09	1.84	57.5[a]	45.8
CH$_4$(g)	0.01	1.09	396.4	397.6
SiH(g)	−0.09	1.52	73.2[a]	74.7
SiH$_4$(g)	−0.04	1.46	304.4	309.5
GeH(g)	0.00	1.59	71.5	73.6
GeH$_4$(g)	0.00	1.54	286.8	276.7
SnH(g)	−0.06	1.79	67.4	74
SnH$_4$(g)	−0.02	1.70	252.0[a]	241.7
NH(g)	0.11	1.04	89.8	85.9
NH$_2$(g)	0.07	1.03	183.8	176.9
NH$_3$(g)	0.05	1.02	280.2	280.3
N$_2$H$_4$(g)	0.07	1.02	411.6	411.6
		1.43		
PH(g)	−0.01	(1.42)	74.7	72.3
PH$_2$(g)	−0.01	(1.42)	149.4	153.5
PH$_3$(g)	−0.01	1.42	228.9	230.2
P$_2$H$_4$(g)	−0.01	(1.42)	356.3	353.8
		(2.20)		
AsH$_3$(g)	0.02	(1.51)	218.4	212.7
SbH$_3$(g)	−0.01	1.71	184.5	184.3
OH(g)	0.19	0.97	103.9	102.4
H$_2$O(g)	0.12	0.96	221.6	221.6
H$_2$O$_2$(g)	0.19	0.97	255.9	260.0
		1.49		
HS(g)	0.07	1.36	85.2	84.6
H$_2$S(g)	0.05	1.33	175.6	175.7
H$_2$Se(g)	0.05	1.47	158.6	146.3
H$_2$Te″(g)	0.00	1.66	126.8	126.8
HF(g)	0.25	0.92	135.5	135.6
HCl(g)	0.16	1.27	103.7	103.3
HBr(g)	0.12	1.41	87.6	87.5
HI″(g)	0.04	1.61	70.4	71.3

[a]Calculated as nonpolar covalent energy only.

in effect it negates any extensive ionicity of the bond. This would cause the bond energy to be essentially that of a hypothetical nonpolar covalent bond instead of the highly polar one. Data testing this hypothesis are also included in Table 7-1.

SPECTROSCOPIC (DIATOMIC) HYDRIDES

Of especial interest here are the monohydrides of elements that normally exhibit a valence greater than one. Table 7-1 lists 12 such compounds. Notice that for the series BH, AlH, GaH, and InH, wherein the outer shell structures of $s^2 p$ would not require any $s-p$ promotional energy, there is no indication of an unusually strong bond as might be expected, all but the InH having calculated bond energies, based on the assumption of cancelled polarity, nearly the same as the experimental value. Similarly, calculated and experimental bond energies are in good agreement for SiH and GeH, although less satisfactory for SnH.

Normal single covalent bonds seem also to describe the bonding in NH and PH, as well as OH and SH. Spectroscopic data are often not capable of producing as accurate experimental bond energies as would be ideal, so small differences may not be significant.

DIVALENT HYDROGEN

Protonic Bridging

Many examples, some extremely important, of seemingly monovalent hydrogen acting in a divalent manner by serving as a bridge between molecules or within a given molecule have long been recognized. The general characteristics of such bridging have been carefully observed and studied, and an extremely complex theory has been developed to explain it. The whole phenomenon is widely known as "hydrogen bonding." My own preference is to call it **protonic bridging**, for reasons to be considered shortly. In an approximate manner, protonic bridging may be described as the result of an electrostatic attraction between the partially bared proton that is the hydrogen nucleus and a pair of electrons on another kind of atom, usually on another molecule. The other kind of atom is usually described as "highly electronegative," but what is meant is *originally* or *initially* highly electronegative and consequently now *possessing substantial negative charge*. Most commonly, protonic bridging involves nitrogen, oxygen, or fluorine.

When studied experimentally, and in consideration of partial charges arising from electronegativity equalization, protonic bridging is revealed to require the following:

(1) A partial positive charge on the hydrogen. Without withdrawal of part of

the hydrogen electron, the hydrogen nucleus is too well shielded to exert a significant positive charge that could attract electrons from elsewhere.

(2) A partial negative charge on the other atom. Without such a negative charge, the other atom would not be able to provide an electron pair to be attracted to the positive hydrogen. A partial negative charge on the other atom means, of course, that initially the other atom must have been relatively high in electronegativity compared with others in its molecule.

(3) The negatively charged atom must be relatively small. It appears, empirically, that protonic bridging does not occur, even when the first two requisites are fully met, if the more negative atom is too large. One assumes that, in order to form a significant bridge to the protonic hydrogen, the electron pair on the partially negative atom must be concentrated within a relatively small region. On larger atoms there is too much space to require such concentration, and the bridging is negligible under ordinary conditions. These conclusions are based on the existence of significant bridging in NH_3, H_2O, and HF, but not in HCl, HBr, or H_2S.

A protonic bridge has also the following characteristics:

(1) It is linear where possible but may be bent when an intramolecular bridge is formed, as required by the geometry of the molecule.

(2) The proton does not sit midway between the two nuclei but remains closest to its original bond and substantially farther away from the bridged atom, causing the bridge to be unsymmetrical.

(3) In effect, a positive hydrogen atom serves as bridge between two pairs of electrons, one on each bridged atom.

(4) The energy of interaction may range from about 1 to 10 kpm.

The data of Table 7-2 reveal that the protonic bridge can be reasonably accurately represented by a simple electrostatic picture.

TABLE 7-2

Estimation of Electrostatic Protonic Bridge Energy in Gases and Liquids

Compound	δ_H	δ_X	R(X–H)	R(X...X)	R(XH...X)	E_{calc}	E_{exp}
NH_3	0.05	−0.16	(1.01)	3.10	(2.09)	1.3	3.8(g), 1.3(c)
H_2O	0.12	−0.25	(0.99)	2.76	(1.77)	5.5	5.0
HF	0.25	−0.25	(0.92)	2.55	(1.63)	12.7[a]	6.8
CH_3OH	0.07	−0.29	(1.07)	(2.76)	(1.69)	5.6	6.0
HCOOH	0.16	−0.21	1.07	2.73	1.66	6.7	7.1
CH_3COOH	0.11	−0.26	(1.07)	2.76	(1.69)	5.6	7.0

[a]If the linear zigzag polymers are planar, the bond angle of 120° permits alternate fluorine atoms to be close enough for about 5–6 kpm repulsion, reducing the calculated net energy by that amount.

Hydridic Bridging

Neutral hydrogen appears incapable of bridging of any kind. When the hydrogen bears a partial negative charge, however, it is capable of bridging two molecules (or two parts of a molecule) if the following requisites are met:

(1) A negative partial charge on hydrogen. Only with an excess of electrons can a hydrogen atom provide electrons for two bonds at once.

(2) A partial positive charge on the other atoms, together with an otherwise vacant orbital on each. The partial positive charge makes the vacant orbital attractive to the electrons of the hydrogen.

This kind of bridging is called **hydridic bridging** to emphasize the distinction from protonic bridging. Other characteristics are:

(1) The bridge is nonlinear.

(2) The hydrogen atom is equidistant from the two bridged atoms.

(3) One pair of electrons appears to be the agent for holding together two other atoms by means of the bridging hydrogen, and therefore the two bonds formed may be regarded as "halfbonds."

(4) The bond energy is from three to ten times larger than for a protonic bridge.

A protonic bridge is a weak, unsymmetrical, linear bridge in which a positive hydrogen attracts two pairs of electrons, each on a partially negative atom. A hydridic bridge is a much stronger, symmetrical, nonlinear bridge in which a negative hydrogen attracts two positively charged atoms, each having a vacant orbital. There is not even a continuous range between the two types, for neutral hydrogen does not bridge.

Bifluoride Ion

Textbooks often cite the bifluoride ion, $F-H-F^-$, as a "typical example of hydrogen bonding." Actually, it seems as atypical as possible. It is linear but symmetrical, the hydrogen lying halfway between the two fluorine nuclei. The bond strength is estimated as several times greater than that of an ordinary protonic bridge. Furthermore, calculation of partial charges shows a *negative* charge on the hydrogen, so that the bridge is actually not protonic at all. Bond energies calculated from this model are in reasonable agreement with estimates made from physical measurements and suggest that the ion involves two half-bonds in a trinuclear system.

THE PERIODICITY OF HYDROGEN CHEMISTRY

The intermediate value of the electronegativity of hydrogen is responsible for an exceptionally wide range in the properties of hydrogen and its binary compounds. Its significance is that hydrogen can react with less electronegative elements, becoming partially negative at their expense, or it can react with more electronegative elements, becoming partially positive.

In general, it is important to recognize that, proportionately, partial charge on hydrogen is much more significant than a similar charge on other atoms, a charge of -1 actually doubling the electron population and a charge of $+1$ corresponding to no electrons at all. When neutral hydrogen itself reacts with a less electronegative element, it plays the role of an oxidizing agent, and when it reacts with a more electronegative element than itself, it serves as reducing agent. The effect of partial charge is to expand these properties.

Negative hydrogen must be an even better reducing agent, potentially, than neutral hydrogen. Positive hydrogen must be an even more active oxidizing agent than neutral hydrogen. Furthermore, there is a tendency for binary compounds of negative hydrogen to interact with binary compounds of positive hydrogen, liberating elemental hydrogen gas and forming a residue of salt containing the positive portion of the compound of negative hydrogen and the negative portion of the compound of positive hydrogen. This reaction is most familiar as the hydrolysis of a hydride, in which the water serves as the binary compound of positive hydrogen.

But hydrolysis is only one of many examples of this kind of reaction. Two principal factors may be recognized. First, negative hydrogen, a reducing agent, is combining with positive hydrogen, an oxidizing agent. Alternatively, negative hydrogen as an electron pair donator is coordinating with positive hydrogen as an electron pair acceptor. Molecular hydrogen is then to be regarded as an oxidation product of negative hydrogen by positive hydrogen, or as a coordination compound uniting hydride ion with a proton. The second and probably more significant factor that promotes the interaction of compounds of negative and positive hydrogen is the fact that if A in AH is less electronegative originally than hydrogen, and B in HB is more electronegative originally than hydrogen, then the most polar bond that can be formed will be between A and B. This will also be the most stable possible bond, contributing to the probably high exothermicity of the reaction.

Another characteristic of hydrogen compounds wherein hydrogen has appreciable charge is their tendency to form complex compounds. The most familiar examples are of course ammonium compounds, in which positive hydrogen has coordinated to an electron pair on a suitable donor, ammonia, forming a complex ammonium ion with positive charge. An opposite example is provided by the complex hydrides such as sodium borohydride, $NaBH_4$. Here the negative hydrogen of NaH acts as donor toward BH_3, in which the hydrogen is also negative but less so, and the boron is both positive and has a vacant orbital to serve as acceptor. BH_3, incidentally, always occurs as diborane (see Fig. 7-1), B_2H_6, in which two BH_2 groups are held together by a double hydridic bridge. In a sense the hydridic bridge is formed to utilize the vacant orbital on boron, but it can be displaced by any electron pair donor that is more effective as a donor than the hydrogen on boron. In other words, more negative hydrogen than that on BH_3 will tend to form stable complexes, but equally negative or less negative or positive hydrogen will offer no advantage and the hydridic bridge will be favored over the complex. Stable

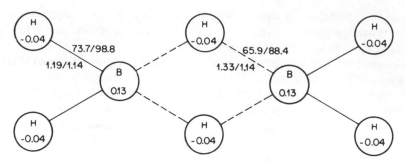

Fig. 7-1 Schematic representation of diborane. Atomization energy: 572.0 kpm, calc.; 573.1 kpm, exp.

complex hydrides are formed when the donor hydrogen is more negative than the hydrogen in the acceptor molecule but generally not otherwise. For example, there exists aluminum borohydride, AlB_3H_{12}, or $Al(BH_4)_3$, but not boron aluminohydride, $B(AlH_4)_3$, for the partial negative charge on hydrogen in boron hydride is smaller than that in aluminum hydride. See Fig. 7-2.

In keeping with its oxidizing tendency, positive hydrogen is of course "protonic" or acidic. There is no direct correlation between positive charge on hydrogen and acidity in aqueous solution, although acidity in general does depend on the ability of the species to provide protons. In aqueous solutions it is observed that some of the least positive hydrogen is most acidic, but this can be understood in terms of the competitive reaction between the solvent, water, and the anion of the acid:

$$HA + H_2O = H_3O^+ + A^-$$

A proton tends to sit on whatever pair of electrons interacts most favorably with it. For example, an electron pair on a highly negative atom interacts more favorably

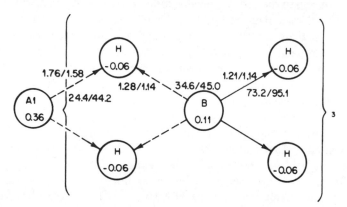

Fig. 7-2 Schematic representation of aluminum borohydride. Atomization energy: 1105.2 kpm, calc.; 1103.7 kpm, exp.

and attracts a proton more strongly. The extent of ionization of a binary hydrogen compound in water solution reflects a competition between two bases, the anion of the acid and the water molecule, for the same protons. It makes little difference which base acquires the proton first. What counts is which base can hold the proton best. If the anion holds the proton best, then the acid dissociation is likely to be very small, and the hydrogen compound is termed a weak acid. But if the water molecule is basic enough to attract protons better than the anion does, the acid molecules will tend to dissociate rather completely in aqueous solution. In most such solutions, the concentration of water molecules is much higher than that of acid molecules or their ions, giving an immediate advantage to water molecules and ensuring that at least some dissociation will probably occur.

In the example of the hydrogen halides, fluoride ion is a much better electron pair donor to a proton than are any of the other halide ions. Consequently, HF (δ_H 0.25) is a relatively weak acid, whereas the others are all strong. Here the leveling effect is observed. Since all three halogen acids, HCl, HBr, and HI (δ_H 0.16, 0.12, and 0.04), are stronger acids than water, they are changed in aqueous solution to the cation of water, H_3O^+, and have the same strength therein. To differentiate among them, it is necessary to test their degree of dissociation within a more acidic solvent, or less basic one, such as glacial acetic acid, wherein the competition between acid anion and solvent molecule is more nearly equal.

The physical state of binary hydrogen compounds depends on the ability of the individual molecules to condense together. Where the hydrogen has a high partial negative charge, its electronic cloud is so very polarizable that "cations" readily condense around it and a crystalline solid resembling ordinary salt results. Consequently the alkali metal hydrides and the alkaline earth metal hydrides are fairly stable, relatively high melting solids of a nonmolecular nature. The condensation energy for the alkali metal hydrides, 40–50 kpm, is comparable to that of the alkali metal halides. Although only lithium hydride is sufficiently stable for melting without decomposition, the others can be dissolved in other molten salts and electrolyzed, yielding hydrogen at the anode as expected for negative hydrogen.

When the negative charge on hydrogen is somewhat less but still significant, association through hydridic bridging seems to be the rule. This is observed in hydrogen compounds of beryllium, magnesium, boron, aluminum, and possibly others. It is not possible beyond Group M3, where there are no longer vacant orbitals to become involved with the pair of electrons of the negative hydrogen. This hydridic bridging results in polymeric condensation in all but the boron hydrides, wherein it tends to be confined to individual molecules, such as diborane, B_2H_6, tetraborane, B_4H_{10}, pentaborane, B_5H_9, and decaborane, $B_{10}H_{14}$. The condensation polymers are thus mostly solids.

When the negative charge on hydrogen is low, but hydridic bridging is impossible because orbital vacancies are lacking, or when the charge on hydrogen is neutral or partially positive, the compounds with hydrogen are all molecular and volatile and mostly gaseous at ordinary temperatures. Protonic bridging causes ammonia, water, and hydrogen fluoride, although all easily volatile, to be much

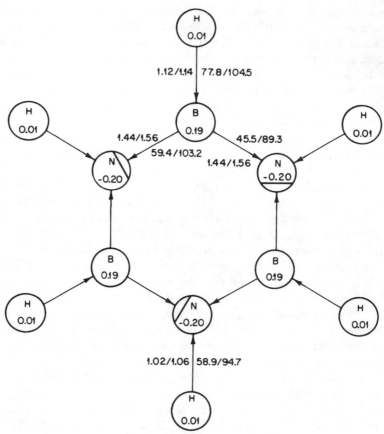

Fig. 7-3 Schematic representation of borazine (all ring bonds are equal). Atomization energy: 1175.1 kpm, calc; 1177.4 kpm, exp.

more strongly associated than otherwise possible and therefore to melt higher and boil higher than any of their congeners. Otherwise there is no appreciable inter-molecular association observed for any of the compounds of hydrogen that are not salts or condensed through hydridic bridging.

In summary, the periodicity of binary hydrogen compounds is very conspic-uous among the major group elements (see Fig. 7-4). The trends from left to right across the periodic table of major groups are from coordinated polymeric solid through hydridic bridged polymer to volatile molecules unassociated except where protonic bridging occurs and therefore mostly gaseous. The trend is one of dimin-ishing reducing power and diminishing electron donating power, toward increasing oxidizing power and increasing acidity, including coordination of the proton. Reactivity toward oxygen and toward water decreases in the same direction. Compounds at one extreme tend to react with compounds at the other extreme to liberate hydrogen. See Table 7-3.

All of these normal characteristics of combined hydrogen are closely correlated with the partial charge on the hydrogen atoms. It is quite evident, and indeed
has been quite evident for at least fifteen years, that the conventional designation
of binary hydrogen compounds as "salt-like" and "covalent" is inadequate and
confusing and internally inconsistent compared to classification according to partial
charge on hydrogen, which gives nearly perfect correspondence with both physical
and chemical properties of the hydrogen compounds. The old argument that the
partial charges on hydrogen are of questionable validity seems thoroughly weakened, not merely by the consistent utility of these values in application to

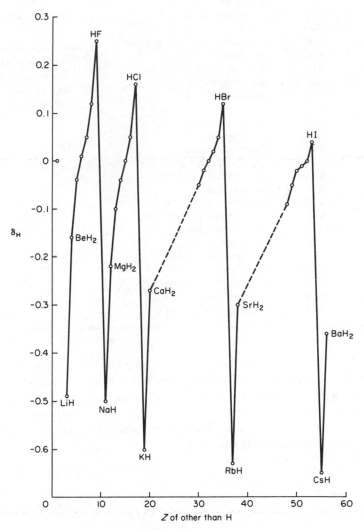

Fig. 7-4 Periodicity of charge on hydrogen.

TABLE 7-3

Partial Charge on Hydrogen and Properties of Binary Hydrogen Compounds

Compound	δ_H	Physical nature	Chemical trends, approx.
HF	0.25	gas, protonic bridging	
H_2O_2	0.19	liquid, protonic bridging	
HCl	0.16	gas, unassociated	
HBr	0.12	gas, unassociated	
H_2O	0.12	liquid, protonic bridging	Increasing
H_2S	0.05	gas, unassociated	acidity,
H_2Se	0.05	gas, unassociated	oxidizing
HI	0.04	gas, unassociated	
NH_3	0.05	gas, protonic bridging in liquid	
AsH_3	0.02	gas, unassociated	Nearly neutral H
CH_4	0.01	gas, unassociated	not very reactive.
H_2Te	0.00	gas, unassociated	May exhibit
GeH_4	0.00	gas, unassociated	intermediate qualities
PH_3	−0.00	gas, unassociated	
SbH_3	−0.01	gas, unassociated	
SnH_4	−0.02	gas, unassociated	
SiH_4	−0.04	gas, unassociated	
B_2H_6	−0.04	gas, dimeric through hydridic bridging	
AlH_3	−0.10	solid, hydridic bridging	
BeH_2	−0.16	solid, hydridic bridging	
MgH_2	−0.22	solid, hydridic bridging?	Increasing
CaH_2	−0.27	coordination polymer (salt)	basicity,
SrH_2	−0.30	coordination polymer	reducing
BaH_2	−0.36	coordination polymer	
LiH	−0.49	coordination polymer	
NaH	−0.50	coordination polymer	
KH	−0.60	coordination polymer	
RbH	−0.63	coordination polymer	
CsH	−0.65	coordination polymer	

understanding the nature of the hydrogen compounds, but also, and perhaps especially, by the excellence of agreement between calculated and experimental bond energies and heats of atomization. The latter applies, of course, only to hydrogen compounds in which the hydrogen does not have high negative charge, which unfortunately appears to invalidate the method of calculating bond energy in a manner not yet quantitatively describable.

EIGHT

The Chemical Behavior
of Oxygen

BONDING BY OXYGEN

In its range of capacity to unite with other atoms, the oxygen atom shows a remarkable versatility. The abundance of oxygen in the crust of the earth has of course ensured that oxides should be an extremely important part of that crust. This is especially true in view of the great stability, characteristic of many oxides, that originates from the high polarity of their bonds, which in turn reflects the very high electronegativity of oxygen itself.

From the electronic configuration of oxygen atoms, 2–6, we recognize easily the presence of two outermost half-filled orbitals corresponding to the two vacancies. These give the capacity to form two covalent bonds or one double bond. In addition, if the two single electrons become paired, this leaves one complete outer orbital vacant and capable of coordinating to a pair of electrons from some donor atom under favorable conditions. In contrast, when oxygen has already formed its two covalent bonds or their equivalent, thus acquiring partial negative charge as the result of its initially higher electronegativity, its lone pair electrons become potentially available for electron donation in coordination. We would not normally predict that an oxygen atom could form a triple bond, but it apparently does in carbon monoxide, CO, by supplying four of the six bonding electrons, carbon supplying only two. Consequently there are examples of oxygen atoms undergoing "normal" covalence by forming two single bonds, one double bond, or one triple bond per atom, or coordinating either as acceptor or donor.

The lone pair bond weakening effect is very large in oxygen, adding further dimensions to its bonding. The oxygen atom may contribute, to the nonpolar covalent portion of its bond energy with another atom, either the O' energy of 33.5, the O'' energy of 68.7, or the O''' energy of 103.9 kpm, depending on conditions. In its single covalent bonds, the value of 33.5 is usually involved, although some exceptions occur. In its double covalent bonds, the value of 68.7 kpm is always used and in its triple covalent bond, the value, 103.9 kpm. However, there are circumstances under which part or all of the lone pair bond weakening effect appears to be removed, even though oxygen is engaged in single bonding. When the other atom possesses outermost orbitals that are not otherwise involved in its bonding, it appears that these may affect the oxygen lone pair electrons in such a manner that they have reduced weakening effects. For example, the outermost d orbitals in silicon in SiO_2 may be considered available to the oxygen lone pairs in this manner, for it is found that the single bond between silicon and oxygen is $Si-O''$ rather than $Si-O'$. As another example, the two lone oxygens (as opposed to hydroxyl oxygens) in sulfuric acid appear to be $S-O'''$ bonds rather than $S-O'$ bonds. Another common example occurs in carboxyl groups, wherein the $C-OH$ bond is $C-O''$ rather than $C-O'$. This seems to depend on the presence of a double-bonded oxygen on the same carbon.

Add to all these possibilities the fact that a complete range of bond polarity is also potentially involved in oxygen bonding, and we have reason to be especially critical of the particular interpretation we may place on the energy of a particular bond to oxygen. Nevertheless, there is usually enough evidence to guide us to a reasonable choice of bond type. We can therefore employ bond energy calculation as a diagnostic tool helping to identify such bonds. Many examples will be provided later in this chapter.

OXYGEN WITH PARTIAL CHARGE

Only in its compounds with fluorine does oxygen acquire partial positive charge. In all other combinations, oxygen becomes partially negative and therefore susceptible to the variations in properties which are associated in general with negative partial charges on atoms. Whatever may be the original properties of the isolated atom, it is certain that any change in the average electron population must influence those properties significantly. In particular, the isolated atom of oxygen has the potential to be strongly oxidizing and to impart acidity to combining hydrogen. Even though there are two lone pairs of electrons in the outermost principal quantum level of an oxygen atom, they are only weakly available through neutral oxygen acting as donor. Acquisition of surplus electronic charge by the oxygen atom must decrease the attraction which the oxygen has for electrons, thus reducing its oxidizing power in whatever compound it may form and increasing its potential as an electron pair donor. By the time the oxygen atom has acquired a fairly high partial negative charge, it must therefore have lost essentially all of its oxidizing power and become a good electron pair donor, which means strongly basic. (It will never become a good reducing agent because of the initially very high electronegativity.)

Combined oxygen bearing small partial negative charge tends to be oxidizing and acidic, changing with increasing negative charge to nonoxidizing (but hardly reducing) and basic.

Accompanying these chemical changes are significant changes in physical properties. With low negative charge on oxygen, the compounds are most likely to be molecular, low melting, and volatile, but as the charge becomes higher, polymeric solids and nonmolecular solids having high melting points and low volatility tend to result. Since oxygen is located at the upper right of the periodic table, we may summarize by saying that the periodicity of compounds of oxygen is well correlated with the partial charge on the combined oxygen. See Table 8-1 and Fig. 8-1.

BOND ENERGIES IN VOLATILE OXIDES

Bond energies or atomization energies have been calculated for 33 binary oxides in the gaseous state, including all such compounds for which data could be obtained. The results, along with some of the basic data, are tabulated in Table 8-2, where each experimental value is also listed for comparison. In many of these compounds, especially when the compound is naturally gaseous at ordinary conditions, the nature of the bonding has long been recognized, and there is little reason to doubt the interpretation upon which the calculation is based. For example, the CO molecule is generally accepted to contain a triple bond, and this appears verified by the calculation based on the assumption of a $C \equiv O'''$ bond.

However, especially for molecules having only high temperature existence, the nature of the bonding cannot necessarily be assumed from the stoichiometry when this does not fit the normal valence. For such molecules the calculation of bond energy serves as a diagnostic tool to be used with discretion, caution, and awareness of the possibility of error. The value of such diagnoses is limited by the uncertainty in some of the experimental bond energies, which most commonly are determined by rather inaccurate spectroscopic estimations. For example, the experimental dissociation energy of BeO gas molecule is about 107 kpm, whereas the value calculated for a single bond, $Be-O'$, is 134 kpm. We cannot overlook the possibility that the difference may be due to error, in which case an admirably ingenious explanation could prove useless.

As will be discussed later in connection with the gaseous halides, the evidence is sparse that promotional energies for s electrons commonly contribute significantly to the value of the bond energy, despite the known higher energy of p orbitals of the same principal quantum level. It appears that such promotional energy is probably taken into account in the evaluation of electronegativity and related properties. On the other hand, Table 8-2 lists all the binary gaseous oxides

TABLE 8-1

Some Properties of Binary Oxides in Order of Decreasing
Negative Charge on Oxygen

Oxide	$-\delta_O$	mp°C	Basicity	Acidity
Cs_2O	0.94	490	strong	none
Rb_2O	0.92	>567	strong	none
K_2O	0.89	>490	strong	none
Na_2O	0.81	920	strong	none
Li_2O	0.80	1727	strong	none
BaO	0.67	2547	strong	none
SrO	0.62	2457	strong	none
CaO	0.57	2587	strong	none
Tl_2O	0.54	300	strong	none
MgO	0.50	2802	medium	none
BeO	0.42	2547	very weak	very weak
SnO	0.37	–	very weak	very weak
PbO	0.36	890	very weak	very weak
CdO	0.32	>1227	very weak	very weak
Al_2O_3	0.31	2027	very weak	very weak
ZnO	0.29	1975	very weak	very weak
HgO	0.27	d	very weak	none
H_2O	0.25	0	very weak	very weak
B_2O_3	0.24	450	none	very weak
SiO_2	0.23	1700	none	very weak
In_2O_3	0.23	–	very weak	very weak
Tl_2O_3	0.21	717	very weak	none
Bi_2O_3	0.20	817	very weak	none
Ga_2O_3	0.19	1725	very weak	very weak
Sb_4O_6	0.18	655	very weak	very weak
P_4O_6	0.18	24	none	weak
PbO_2	0.18	d	very weak	very weak
SnO_2	0.17	1927	very weak	very weak
CO	0.16	–	–	–
Bi_2O_5	0.15	d	none	weak
GeO_2	0.13	1116	very weak	very weak
P_4O_{10}	0.13	–	none	medium
TeO_2	0.13	477	very weak	very weak
As_4O_6	0.12	315	very weak	very weak
CO_2	0.11	−57(P)	none	weak
N_2O	0.10	–	–	–
As_2O_5	0.09	d	none	weak
I_2O_5	0.09	d 300	none	strong
SO_2	0.08	−72.5	none	weak
NO	0.08	low	–	–
SeO_2	0.07	340	none	weak
SO_3	0.06	17, 30	none	strong
SeO_3	0.06	118	none	strong
N_2O_5	0.05	>33	none	strong
Cl_2O	0.04	−116	none	very weak
Cl_2O_7	0.01	−91.5	none	very strong

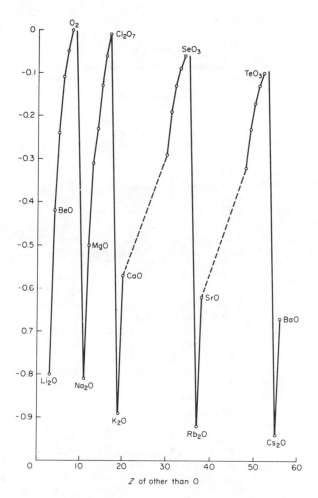

Fig. 8-1 Periodicity of charge on oxygen.

of Group M2, and for each compound the calculated energy for a single bond is substantially higher than the experimental value. The differences are: BeO 27, MgO 34, CaO 32, SrO 22, and BaO 3 kpm. The last value is probably well within the experimental error, but the first four might represent energy lost in promoting one of the outermost s pair of electrons to a p orbital so that the atom can then engage in covalence.

It is interesting that these data give no real hint of any bond multiplicity, suggesting thereby the inability of M2 atoms to use both bonding orbitals in joining to the same other atom. In other words, it appears most likely that all these molecules are held together by single polar covalent bonds. Whether these are coordination bonds (requiring no promotion of an s electron from the pair) with

TABLE 8-2

Atomization of Gaseous Oxides

Oxide	$-\delta_O$	R_0	E_{calc}	E_{exp}	Bond type
HO	0.19	0.97	103.9	102.4	H–O'
H_2O	0.25	0.96	221.6	221.6	H–O'
H_2O_2	0.19	0.97	252.1	256.0	H–O', O'–O'
		1.49			
BeO	0.42	1.33	134.0	107	Be–O'
MgO	0.50	1.75	115.1	81	Mg–O'
CaO	0.57	1.82	123.5	93	Ca–O'
SrO	0.60	1.92	121.8	100	Sr–O'
BaO	0.67	1.94	127.9	131	Ba–O'
BO	· 0.27	1.20	179.8	188, 192	B–O''', B=O''
B_2O_3	0.23	1.24	653.6	649.5	B=O'', B–O'''
		1.36			
AlO	0.38	1.62	119.6	117	Al–O''
GaO	0.23	1.74	92.8	91	Ga–O''
CO	0.16	1.13	260	258	C≡O'''
CO_2	0.11	1.42	384.6	384.6	C=O''
SiO	0.29	1.51	194.0	184, 192	Si=O'''
SiO_2	0.20	1.54	302.3	304.3	Si–O''', Si=O''
GeO	0.19	1.62	161.2	157, 161	Ge=O'''
SnO	0.37	1.83	129.6	127	Sn–O'', Sn=O''
PbO	0.36	1.92	89.4	89	Pb–O''
N_2O	0.10	1.13	264.7	266.1	N'''–'''N''=O''
		1.19			
NO	0.08	1.15	151.8	151.0	N''=O''
NO_2	0.05	1.19	226.8	224.2	N'–O'', N''=O''
PO	0.21	1.47	117.9	120.4, 125	P–O'''
P_4O_6	0.17	1.65	1120.8	1187.5	P–O''
P_4O_{10}	0.12	1.62	1595.8	1588.7	6 P'–O', 6 P'–O'',
		1.39			4 P'''–O'''
BiO	0.24	1.93	80.2	82	Bi–O''
O_2	0	1.21	119.2	119.2	O''=O''
O_3	0	1.28	149.3	143.5	O'–O', O''=O''
			139.6		O'–O''', O''–O''
SO	0.12	1.48	122.6	124.7	S–O''', S''=O''
SO_2	0.08	1.43	255.6	256.7	S–O''', S''=O''
SO_3	0.06	1.43	342.3	340.0	2 S''–O'', S''=O''
SeO	0.11	1.64	106.0	101	Se–O''', Se''=O''
TeO	0.19	1.83	88.4	91	Te–O'''
ClO	0.03	1.55	53.1	64.5	Cl–O'
			73.2		Cl–O''
BrO	0.07	1.67	54.2	56	Br–O'

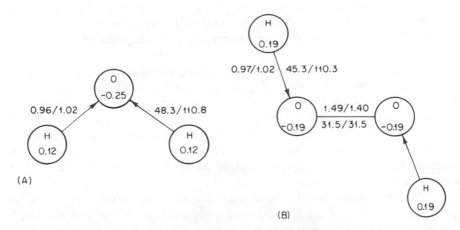

Fig. 8-2 Schematic representation of (A) water and (B) hydrogen peroxide.

oxygen acceptor or diradical molecules would not be established by the calculation. One final observation: Using the same parameters for radii of charged atoms that are successful in the M2 halides, one calculates bond lengths in the M2 gaseous oxides that are significantly longer than the experimental bond lengths, the latter being presumably highly accurate.

From the bond energy calculation method, the bond in gaseous BO appears to resemble the bond in SO, SO_2, and SeO in being an average of an ordinary double bond and a coordination bond to oxygen acceptor in which the weakening effect in oxygen has disappeared: B=O″, B–O‴. But the calculated energy of 180 is somewhat lower than the experimental values of 188 and 192, which suggests that any conclusions about the nature of the bond must be tentative. In the heavier counterparts, AlO and GaO, the bond energy determined experimentally appears to be adequately represented by single bonds in which the weakening effect in oxygen is half removed: Al–O″, Ga–O″. In the "normal" oxide, B_2O_3, in the gas phase the bonding is well represented as involving two B–O″ bonds and two B=O″ bonds: O=B–O–B=O. This is a reasonable interpretation. The example resembles the carboxylic acid group, in which oxygen singly bonded to carbonyl carbon also shows partial removal of the weakening effect by the use of the O″ energy instead of the usual O′ energy for the oxygen single bond.

The data on M4 oxides in the gaseous state are preliminary to a discussion of the marked differences between the oxides of carbon and those of silicon. In carbon monoxide the presumption that oxygen contributes four electrons and carbon only two to the triple bond appears to disclose no difference between this bond and a triple bond wherein each atom provides three electrons. A very low experimental dipole moment of CO suggests that the polarity expected from the electronegativity difference between carbon and oxygen is essentially cancelled by the carbon being "negative" at the expense of the oxygen, which is left "positive."

However, the ionic contribution to the total bond energy is very substantial and apparently exactly as expected. Once again the subtleties of electronic arrangement involved in determining dipole moment appear beyond the scope of simple understanding.

Numerous data suggest that multiple bonding is inhibited by difference in size of the atoms involved. For the analogues of carbon we see evidence that this may be so. Thus the interpretation of bond type as determined from energy calculations is $C\equiv O'''$, $Si=O'''$, $Ge=O'''$, average of $Sn-O''$ and $Sn=O''$, and $Pb-O''$. The agreement between experimental and calculated bond energies is quite satisfactory when these bond types are assumed. However, as always, we must recognize the empirical nature of these explanations at the present time, realizing that errors may be concealed or unrecognized.

One of the noteworthy properties of carbon monoxide is its failure to polymerize, despite the usual tendency of triple bonds to react in that manner. It can be shown that the monomer has greatest stability. In particular the dimer, $O=C=C=O$, would be less table because the $C=C$ double bond, weakened by the presence of partial positive charges on each carbon atom, would not compensate for the change in the CO bonds from triple to double under the circumstances of the lone pair weakening effect.

It is interesting that SiO, but never CO, rapidly disproportionates to form the solid dioxide and free element. Actually carbon monoxide is about 20 kpm less stable than its equivalent in solid graphite and gaseous carbon dioxide, so that it would be expected to disproportionate easily and exothermically. If it were not for the problem of the immediate coating of a solid catalyst with carbon, this might in fact present an approach to the problem of removing carbon monoxide from fuel combustion products. Possibly because of the indicated reduced multiplicity in silicon monoxide, the molecule possesses properties of a free radical, being able to undergo interaction with other molecules very easily (see Fig. 8-3). The carbon monoxide molecule does not have these properties and furthermore is more strongly bonded, requiring a higher activation energy by far for the disproportionation reaction. In addition, the disproportionation of silicon monoxide is much more highly exothermic, in keeping with the much higher stability of solid silicon dioxide compared with carbon dioxide gas.

In reference to the diminished tendency toward multiplicity with increasing size difference between the atoms, it is of interest that silicon dioxide gas appears unlike carbon dioxide gas. The bonds, in fact, are like those in BO, SO, SO_2, and elsewhere: $Si-O'''$, $Si=O''$ average. This fact alone favors a greater probability of condensation to a polymeric solid.

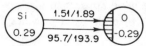

Fig. 8-3 Schematic representation of silicon monoxide. Atomization energy: 193.9 kpm, calc.; 192.3 kpm, exp.

We move now to the realm of nonmetal oxides in which polymerization is relatively weak or does not occur. The first problem, and one that is easily solved, is the extraordinary difference between the oxides of nitrogen and those of the heavier M5 elements. The oxides of nitrogen are multiply bonded monomeric, in contrast to the solid, singly bonded oxides of phosphorus and the heavier elements. These differences arise from two principal factors. One is the lone pair weakening effect, which is much greater in nitrogen and oxygen than in their heavier congeners. This effect favors multiple bonding, wherein part or all of the weakening is removed, over single bonding, where the weakening is in full effect. The other is that because nitrogen is much more electronegative than any other M5 element, it is closer to oxygen in that respect and therefore forms much less polar bonds with oxygen. Remember that the **lone pair weakening effect** applies solely to the nonpolar covalent contribution to the bond energy and therefore **is more important the less polar the bond.**

The first member of the series, nitrous oxide, can be accounted for very accurately, with respect to atomization energy, by assuming that the central nitrogen in the NNO grouping forms a normal double bond with the oxygen, $N''=O''$. It is then incapable of a triple bond to the terminal nitrogen atom but does the best it can: $N'''-N'''$. This may seem a farfetched simplification of the usual complex system in resonance by which N_2O is described, but perhaps it only supplies quantification to the resonance picture in a simple manner. Actually the dissociation energy of the NO bond should be only about 40 kpm, since removal of an oxygen atom would permit formation of the very stable N_2 molecule. As will be discussed in a later chapter, the bond dissociation energy in this case differs markedly from the contributing bond energy.

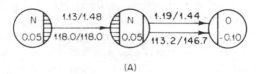

(A)

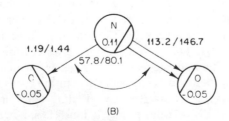

(B)

Fig. 8-4 Schematic representation of (A) nitrous oxide and (B) nitrogen dioxide (both bonds are equal). (A) Atomization energy: 264.7 kpm, calc.; 266.1 kpm, exp. (B) Atomization energy: 226.8 kpm, calc.; 224.7 kpm, exp.

Practically every authority on molecular orbital theory considers the nitric oxide molecule NO to be adequately described as involving a bond order of 2.5, corresponding to a total of six bonding electrons and one antibonding electron. It is therefore especially interesting that the calculation of bond energy gives excellent agreement with the experimental value when the bond is assumed to be simply a double bond, $N''=O''$. According to this bond picture, however, there must be a single outermost electron somewhere in the molecule which should be capable of causing dimerization readily, which nitric oxide does not do. Further study is certainly needed.

If an oxygen atom is substituted for the terminal nitrogen atom in nitrous oxide, the result is NO_2. We might describe the nitrogen as joining to the first oxygen by an ordinary double bond, $N''=O''$, but this would not permit the nitrogen to join in the same way to the second oxygen. If we describe the bonding as $N'-O''$, $N''=O''$ (meaning an average of the two), the atomization energy is in good agreement with the experimental value.

The bonding in phosphorus (III) oxide consists of 12 P–O bonds that appear to be approximately but not accurately represented by $P-O''$. The experimental value of the atomization energy is significantly higher than the calculated value. In part, this would be compensated for by assuming $P'''-O''$ bonds, which would bring the energy to 1176 kpm, compared to the experimental value of 1188 kpm. But without better understanding of the "lone pair weakening effect," there appears to be no justification for treating the bonds in this manner. Normally, when the weakening effect on a polar covalent single bond is reduced, the reduction is on the contribution made by the more electronegative atom only, in the present example, oxygen.

The bonding in phosphorus (V) oxide is equally puzzling. To approach the experimental value of 1589 kpm, it seems necessary to add the energies of six $P'-O'$ bonds, six $P'-O''$ bonds, and four $P'''-O'''$ bonds, totaling 1596 kpm. The four very short bonds at the corners of the molecular tetrahedron do appear to be coordination bonds in which oxygen is acceptor.

Strangely ozone, O_3, appears difficult to classify. A reasonable assumption would be that the central oxygen forms an ordinary double bond to one other oxygen, and the second oxygen coordinates as acceptor to a lone pair on the central oxygen. This would be accompanied by averaging to equal bonds, of course. But the energy calculation on this basis is too high, 149.3 compared to 143.5 for the experimental value. An alternative is that the coordination bond might be $O'''-O'''$ and the other bond a single bond $O''-O''$, which gives too low a value, 139.6, and seems less reasonable. The bond angle of $120°$ suggests one lone pair on the central oxygen not engaged in bonding, in agreement with the first model. Whatever the cause of the discrepancy, the average contributing bond energy is 71.8 kpm, experimental. The bond dissociation energy is much less, however, because of the increase in bond strength in the residue, O_2, to 119.2 kpm. This is an increase of 47.4, which means that the bond dissociation energy for the first oxygen would be reduced by that amount, to $71.8 - 47.4 = 24.4$ kpm. The high degree of reactivity

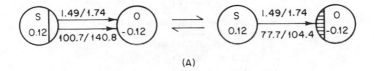

(A)

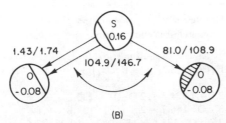

(B)

Fig. 8-5 Schematic representation of SO and SO$_2$ (A) Sulfur monoxide (bond is an average of the two). Atomization energy: 122.6 kpm, calc.; 124.7 kpm, exp. (B) Sulfur dioxide (bonds are equal). Atomization energy: 255.6 kpm, calc.; 256.7 kpm, exp.

of ozone is thus explained in terms of the ease of splitting off an oxygen atom resulting from the stability of the residual O$_2$ molecule.

One of the questions that should come to everyone's mind in the process of considering the reaction between sulfur and oxygen is why, when both elements are clearly divalent, they do not unite to form sulfur monoxide, SO. This molecule has indeed been detected and measured in the gas phase. It is found that the bond can be accurately represented as a blend or average of an ordinary double bond, $S''=O''$, and a coordination bond with oxygen acceptor, $S'-O'''$. If this were an isolated example, one would have more reason to be skeptical, but this type of bonding

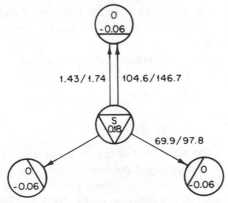

Fig. 8-6 Schematic representation of sulfur trioxide (all bonds equal). Atomization energy: 342.3 kpm, calc.; 340.0 kpm, exp.

TABLE 8-3

Summary of Hydration of Sulfur Oxides

Sulfur Trioxide

$$H_2O(g) + SO_3(g) \rightarrow H_2SO_4(g)$$

2 O–H 221.6 → 2 O–H	221.2	+0.4
S″=O″ 146.7 → 2 S′–O′	137.8	+8.9
S′–O‴ 108.9 → S′–O‴	109.6	−0.7

Net change = heat of hydration   +8.6
Estimated free energy change   +4 kpm

Sulfur Trioxide

$$H_2O(g) + SO_3(g) \rightarrow H_2SO_4(g)$$

2 O–H 221.6 → 2 O–H	223.2	− 1.6
S″=O″ 146.7 → 2 S′–O′	137.8	+ 8.9
2 S″–O″ 195.6 → 2 S′–O‴	219.2	−23.6

Net change = heat of hydration − 16.3
Estimated free energy change − 12 kpm

appears to occur in several places where an ordinary double bond appears inhibited by a difference in size between bonded atoms or by other factors as in BO.

The compound S_2O, also a highly unstable-seeming species, illustrates this point. The compound consists of bent molecules, SSO, and with one double bond and one coordination bond. But the double bond is not between the sulfur and oxygen, as one might be inclined to expect. It appears to be between the two sulfur atoms, which are of course the same size. Then oxygen is attached by a coordination bond, $S'–O'''$. The reason both compounds, SO and S_2O, are little known is that they very readily disproportionate into sulfur and sulfur dioxide. The sulfur to oxygen bonds in SO_2 are essentially the same as in SO, so without changing the number of such bonds it is possible to form solid sulfur or S_8 rings in vapor phase. This provides enough extra energy to make the disproportionation worthwhile.

Notice that when oxygen unites with the larger tellurium atom, no degree of multiplicity is evidenced.

Results for ClO appear uncertain, but for BrO it seems quite clear that the bond is an ordinary, fully weakened single covalent bond.

SOLID OXIDES

Binary oxides of the active metals appear to be described with approximate accuracy in terms of the coordinated polymeric model for nonmolecular solids. It has been observed empirically that the normal procedures for calculating atomization energies of nonmolecular solids containing divalent "anions" invariably produce values much too high, but that these can be corrected by multiplying the ionic energy contribution by the factor, for oxides, of 0.63. It has been proposed that the need for this correction may arise from an unrealistic evaluation of the repulsion coefficient k. If this is evaluated from the experimental compressibility of an oxide crystal, and calculations are based upon the assumption that oxide ions are present, and if in fact the oxygen atoms in the compound do not even closely approach the $O^=$ ionic condition, then perhaps this would be the source of a large error in k. Certainly this calls for additional study and investigation. Table 8-4 lists 10 solid oxides for which reasonably accurate atomization energies can be calculated if this standard factor of 0.63 is used. The same phenomenon, with different correction factors, is observed for other chalcide atoms, notably sulfur and selenium.

The factor n is introduced into the calculation of the covalent energy contribution in nonmolecular solids because of the coordination of atoms such that one may presume all outermost electrons to be involved equally in the bonding. The value of 4 is the common one, and the expected one, since there are normally eight electrons in the outermost levels per formula unit. Use of this factor gives an atomization energy for lithium oxide, Li_2O, that is much too low but corrected by change of the factor to 6. An adequate justification or explanation cannot be supplied at this time.

In general, it is useful to note that the covalent energy contribution to the total atomization energy, labeled $t_c E_c$ in Table 8-4, is very substantial in all these oxides. The usual picture of such oxides is as assemblages of ions. It has long been

TABLE 8-4
Atomization of Solid Oxides[a]

Oxide	n	$-\delta_O$	M	k	R_o	$t_c E_c$	E_{calc}	E_{exp}
Li_2O	6	0.80	5.04	0.83	2.00	108.7	280.3	279.5
Na_2O	4	0.80	5.04	0.85	2.41	54.9	203.6	210.8
K_2O	4	0.84	5.04	0.88	2.79	46.4	186.0	189.2
Rb_2O	4	0.86	5.04	0.88	2.92	45.6	182.2	177.8
Cs_2O	4	0.90	4.38	0.89	3.05	44.3	172.6	173.0
BeO	4	0.42	6.37	0.85	1.60	132.3	280.9	281.0
MgO	4	0.50	1.75	0.82	2.10	101.0	243.9	248.9
CaO	4	0.57	1.75	0.85	2.40	97.9	248.3	253.9
SrO	4	0.60	1.75	0.86	2.57	93.8	240.8	239.8
BaO	4	0.67	1.75	0.86	2.76	74.1	229.2	235.0
ZnO	–	0.27	–	–	1.95	102.0	173.8	173.8
CdO	–	0.32	–	–	2.35	–	146.1	147.5
$(CO_2)_x$	–	0.11	–	–	1.42	181.6	340.4	–
$(SiO_2)_x$	–	0.20	–	–	1.61	199.2	444.8	445.8
$(GeO_2)_x$	–	0.12	–	–	1.89	232.6	341.4	340.9
$(SnO_2)_x$	6	0.18	2.41	0.85	2.05	197.4	347.9	330.0

[a]Oxide factor for $E_i = 0.63$ used for all coordination polymers (where n is given).

recognized that formation of an oxide ion, $O^=$, is an endothermic process requiring about 170 kpm and, of course, that removal of electrons from the metal atoms requires substantial ionization energies. However, it has been rationalized that the electrostatic interaction among oppositely charged ions is more than enough to compensate for the energy required to produce the ions. This is quite true. For example, more energy is gained by bringing magnesium ions together with oxide ions than it costs to create them. However, it has never seemed logically acceptable to this author that an oxide ion that could rid itself of electrons exothermically could be placed within a crystalline environment wherein it is surrounded on all sides and in direct contact with six or eight electron-hungry cations and remain recognizable as an $O^=$ ion. I would be perfectly willing to accept the description of a solid oxide as what one would form by placing together appropriate numbers of cations and oxide ions, but then to picture the solid crystal as consisting of ions appears totally unrealistic. The magnitude of covalent interaction in the oxides tabulated is indicated to be never less, and often much more, than 25% of the total atomization energy.

As described in Chapter Six, not all simple solids of a nonmolecular nature conform to the coordinated polymeric model. As indicated in the table, some, such as oxides of zinc and cadmium, appear to involve electrostatic condensation. This is an area needing much further development.

In Table 8-4 are presented data for solid carbon dioxide, the hypothetical polymer in which each carbon atom is joined to four oxygen atoms through polar single covalent bonds (see also Fig. 8-8). It will be noted that the atomization energy of such a polymer would be 340.4, compared to 384.6 for the gaseous monomer. The reason for the difference in that direction is that the weakening

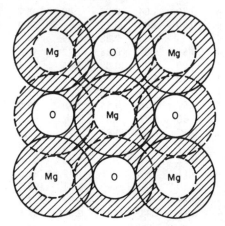

Fig. 8-7 Covalent and ionic magnesium oxide. Solid lines represent nonpolar covalent radii. Broken lines represent ionic radii. Striated area represents region occupied by valence electrons.

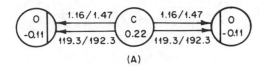

(A)

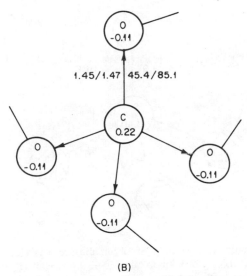

(B)

Fig. 8-8 Schematic representation of carbon dioxide. (A) Monomer, atomization energy: 384.6 kpm, calc.; 384.6 kpm, exp. (B) Polymer, atomization energy: 340.4 kpm, calc.; heat of polymerization: +44.2 kpm.

effect on the oxygen homonuclear energy is partly removed when the oxygen forms a double bond but not in the single bonds. Therefore two double bonds of oxygen to carbon are more stable than four single bonds. The enthalpy difference of 44.2 kpm suggests that this represents the endothermicity of polymerization of gaseous carbon dioxide. Polymerization would, of course, result also in a decrease in entropy, so that the free energy of polymerization would be even more positive than the enthalpy.

Silicon atoms, on the other hand, differ from carbon atoms in being much less electronegative and therefore having higher positive charge in the oxide. In addition, positive charge contributes to the availability of outermost d orbitals on silicon, which are not available at all on carbon. The result seems to be that the lone pair weakening effect is partly reduced when oxygen forms even a single bond to silicon, using the E'' energy of 68.7 instead of the E' energy of only 33.5 kpm as in combination with carbon. There is no question here, in silicon dioxide, that two double bonds are not as stable as four single bonds. Therefore SiO_2 is a highly stable polymeric solid having no tendency to emulate CO_2 by becoming a gaseous

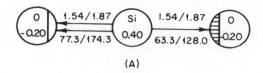

(A)

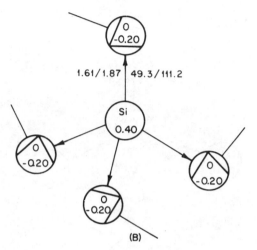

(B)

Fig. 8-9 Schematic representation of silicon dioxide. (A) Monomer (both bonds are equal). Atomization energy: 302.3 kpm, calc.; 304.3 kpm, exp. (B) Polymer. Atomization energy: 444.8 kpm, calc.; 445.8 kpm, exp.

monomer (see Fig. 8-9). Also remember, from the earlier discussion of SiO_2 gas, that even in this molecule there are no conventional double bonds but that the bonding is of a weaker type. Thus is provided a very satisfactory quantitative answer to the question of long standing about the striking physical differences between CO_2 and SiO_2.

HYDROXIDES, OXYACIDS

Atomization energy data for a number of real and hypothetical hydroxy compounds are given in Table 8-5.

As has been discussed in Chapter Seven, hydrogen bearing appreciable partial negative charge would be expected to be more highly polarizable than any other possible species. It appears that the polarization effects almost completely cancel the polarity effects in bonds to negative hydrogen, so that the bond energy can be described with approximate accuracy as a nonpolar covalent energy. The hydrogen

TABLE 8-5

Atomization of Hydroxy Compounds

Compound (gas)	Bond	$-\delta_O$ (or t_i)	R_0	Bond E, calc.	Atomization energy Calc.	Atomization energy Exp.
LiOH	Li–O	0.60	1.82	145.4	206.9	208
	H–O	0.60	0.98	61.5[a]		
HOBO	H–O	0.23	0.96	112.9	442.9	440.1
	B–O″	0.23	1.38	126.8		
	B=O″	0.23	1.24	203.2		
B(OH)$_3$	H–O	0.24	0.96	113.1	723.3	728.2
	B–O″ ⎱	0.24	1.38	128.0	(705)	
	B–O‴ ⎰				(742)	
BF$_2$OH	H–O	0.16	0.96	113.0	541.4	544.9
	B–O″	0.16	1.38	121.7	(+ protonic	
	B–F‴′	0.34 (t_i)	1.30	153.3	bridge)	
BF(OH)$_2$	H–O	0.21	0.96	112.0	619.2	626.5
	B–O″	0.21	1.38	121.7	(+ 2 protonic	
	B–F‴′	0.33 (t_i)	1.30	151.8	bridges)	
B$_3$O$_3$(OH)$_3$	H–O	0.23	0.96	112.9	1450.2	1468.2
	B–O″	0.23	1.36	123.5	(+ 3 protonic bridges)	
H$_2$CO$_3$	H–O	0.18	0.95	112.4	578.4	
	C–O′	0.18	1.38	87.6		
	C=O″	0.18	1.25	178.4		
HNO$_2$	H–O	0.14	0.96	112.4	309.6	303.2
	N′–O′	0.14	1.44	51.7		
	N″=O″	0.14	1.20	145.5		
HNO$_3$	H–O	0.11	0.96	111.5	370.7	376.2
	N′–O′	0.11	1.44	51.7	(+ protonic	
	N′–O′	0.11	1.20	62.0	bridge)	
	N″=O″	0.11	1.20	145.5		
H$_3$NO$_4$	H–O	0.16	0.96	110.6	565.4	
	N′–O′	0.16	1.44	58.4		
H$_2$SO$_3$	H–O	0.17	0.97	110.6	468.6	
	S′–O′	0.17	1.53	68.9		
	S′–O‴′	0.17	1.42	109.6		
H$_2$SO$_4$	H–O	0.15	0.97	111.6	580.2	586.2
	S′–O′	0.15	1.53	68.9	(+ protonic	
	S′–O‴′	0.15	1.42	109.6	bridging)	
N(OH)$_3$	H–O	0.19	0.96	114.2	517.9	
	N–O	0.19	1.44	58.4		
HClO$_4$	H–O	0.08	0.95	111.5	336.1	
	Cl–OH	0.08	1.64	50.3		
	Cl–O′	0.08	1.42	174.3		

[a] Assuming nonpolar bond.

in LiOH is calculated to have a partial charge of −0.30, but by considering the O—H energy to be that of a nonpolar bond, 61.5 kpm, the total atomization energy of the molecule is correctly determined. It is of interest that the O—H bond length is only 0.98 Å, indicating also that the hydrogen makes only minor contribution to this value.

It has long been recognized that two hydroxyl groups attached to the same carbon atom tend to split off water very easily. This is presumably kinetically reasonable since the hydroxyl groups are close together, making at least an intra-molecular reaction feasible. Thermodynamically it is reasonable also, as we can calculate. One example is the instability of carbonic acid, $(HO)_2CO$. As shown in the table, the calculated atomization energy of this species is about 578 kpm. In comparison, the atomization energies of carbon dioxide and water, the decomposi-tion products, are 221.6 and 384.6, totaling 606 kpm. Thus the enthalpy of decomposition would be about 578 − 606 = −28 kpm. We may assume that the carbonic acid would be liquid and the water liquid, with both exhibiting about the same degree of protonic bridging. If so, then the decomposition would increase the entropy by releasing carbon dioxide gas, which would cause the free energy of the decomposition to be even more negative than the enthalpy. The absence of H_2CO_3 as a stable species is therefore the result of a kinetically convenient means of rearrangement of atoms to increase both bond strength and entropy.

Similar arguments apply to such examples as the nonexistence of nitrogen trihydroxide, or orthocarbonic acid, or orthonitric acid, H_3NO_4. From the table, we see that the atomization energy of nitrous acid is 303 kpm, whereas that of the trihydroxide (orthonitrous acid) is 518 kpm. Formation of a molecule of water from the orthonitrous acid, with 221.6 kpm atomization energy, added to the 303 of the metaacid, gives 525 kpm, indicating a little greater stability of the decompo-sition products. For metanitric acid and water the atomization energies of 376 and

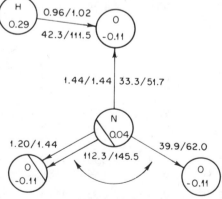

Fig. 8-10 Schematic representation of nitric acid. Atomization energy: 370.7 kpm, calc.; 376.2 kpm, exp. (Internal protonic bridging.)

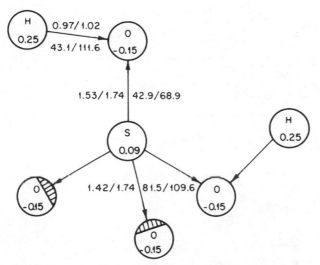

Fig. 8-11 Schematic representation of sulfuric acid. Atomization energy: 580.2 kpm, calc.; 586.2 kpm, exp. (Internal protonic bridging?)

222 total 598 kpm, compared to 565 calculated for the orthonitric acid. Thus the absence of orthonitric acid, corresponding to orthophosphoric acid, is a reasonable phenomenon. In contrast, orthophosphoric acid is more stable than its decomposition products, so the metaacid hydrates but the orthoacid does not dehydrate.

It can similarly be shown that sulfurous acid is unstable and sulfuric acid stable toward dehydration. (See Table 8-3.) The application of this tool can give us insights concerning the nonexistence and existence of certain combinations of atoms and rationalize many common chemical phenomena which heretofore have been unduly puzzling.

NINE

Halide Chemistry

PARTIAL CHARGE ON HALOGEN AND HALIDE PROPERTIES

Physical Properties

For two closely related reasons, chemical combinations of the halogens with other nonmetals tend to be molecular and volatile. First, other nonmetals characteristically have more electrons than vacancies in their outermost shells and therefore, when they have formed halides, have no opportunity to condense further because of the lack of vacant orbitals. Second, nonmetals as a class are high in electronegativity, and bonds to halogen cannot be very polar. This factor also minimizes the tendency toward further condensation. But metal atoms do provide outer vacant orbitals and also form more polar bonds. Both permit condensation to higher aggregates, usually nonmolecular solids. In general, the partial charge on the halogen atom serves as a satisfactory index of the ability of the halide to condense further.

Metal halides can, however, be molecular and volatile if the number of halogen atoms per metal atom is sufficient to surround the metal atom filling all vacancies or preventing them from further use. In such compounds the competition among halogen atoms will prevent any one of them from becoming very negative anyway, so that the partial charge remains a reliable index in an approximate manner.

The alkali metal halides are the compounds having the highest partial negative charge·on halogen, for two reasons. One is that the electronegativity of the metal is very low. The other is that the valence is low so there can be no competition among halogen atoms. Both the ionic and the coordinated covalent models of these salts appear to furnish adequate quantitative descriptions of the bonding, but the latter model is so much more widely applicable that it appears superior.

Among the several kinds of halide, the properties indicating strength of aggregation, such as melting point, boiling point, and atomization energy, tend to decrease in the order, fluoride, chloride, bromide, and iodide.

Chemical Properties

There are two principal characteristics of halides that tend to govern their chemistry. One is oxidizing power, which is diminished to the extent that the halogen acquires partial negative charge but which may remain quite significant when this partial negative charge is still small. As the partial charge on halogen increases, the oxidizing power of the halogen decreases until it is essentially nonexistent in the alkali metal halides. Thus in a series of polyhalides of a given element, the compound containing the greatest number of halogen atoms per atom of the other element is the strongest oxidizing agent, and the oxidizing power diminishes toward the lower oxidation states of the other element. As will be considered in detail, there is also the reactivity of the halide to be taken into account in addition to partial charge on halogen. Polyhalides in which some of the halogen atoms are held by half-bonds are highly reactive because of the ease with which atomic halogen can be released.

The second general chemical characteristic of halides is their tendency toward hydrolysis. In reaction with water, the hydrogen halide tends to be formed, and the hydroxyl group replaces the halogen on the original halide. The extent of hydrolysis to be expected depends largely on the nature of the hydroxide so formed. If the hydroxide is a strong base, then no hydrolysis occurs. If the hydroxide is a weak base, partial hydrolysis will occur. When the hydroxide has no basic properties at all, being exclusively acidic, then there is no way in which the hydroxide ion can separate again to be replaced by the halide ion, so hydrolysis is complete and irreversible. Since, as indicated in Chapter Eight, the partial charge on oxygen strongly influences the behavior of oxides and hydroxides as acids and bases, the partial charge on halogen will also be an index of the tendency for hydrolysis to occur. In general, complete hydrolysis is expected of halides in which the partial charge on halogen is only slightly negative, and the hydrolysis tends to be reversible and finally nonexistent as the partial charge on halogen increases to a high negative value.

Among the several halides the greatest resistance to hydrolysis is offered by the fluorides. This is related to the higher base strength of the fluoride ion.

The use of outer lone pairs on halogen in halide, for coordination as electron pair donors, is of course greatly aided by high negative charge on the halogen and becomes negligibly small when the negative charge on the halogen is small.

Periodicity of Halides

Because of the periodicity of electronegativity and other related properties, the halides themselves exhibit marked periodicity according to the periodic table. From left to right across the major groups, the halides tend to change from stable, high-melting, nonvolatile solids toward halides polymeric through halide bridging to molecular and volatile compounds. At the same time, the electron pair donating ability diminishes, the tendency toward hydrolysis increases, and the halogenating or oxidizing ability increases.

The relationship of these changes to changes in partial charge on halogen is illustrated by Table 9-1, which lists only binary chlorides in order of decreasing negative charge on chlorine. See also Figure 9-1. The data for other halides would be very similar.

TABLE 9-1
Chloride Properties and Partial Charge on Chlorine

Compound	$-\delta_{Cl}$	mp, °C	bp, °C	Hydrolysis	Oxidizing or chlorinating ability
CsCl	0.82	918	1573	none	none
RbCl	0.78	990	1654	none	none
KCl	0.76	1045	1680	none	none
NaCl	0.67	1081	1738	none	none
LiCl	0.66	883	1655	none	none
$BaCl_2$	0.49	962	–	none	none
$SrCl_2$	0.44	875	–	none	none
TlCl	0.41	429	806	none	none
$CaCl_2$	0.40	782	–	none	none
$MgCl_2$	0.34	987	1691	slight	none
$BeCl_2$	0.28	–	547	some	none
$SnCl_2$	0.24	247	–	some	none
$PbCl_2$	0.23	498	954	some	none
$CdCl_2$	0.21	841	1253	some	none
$AlCl_3$	0.19	192	180s	extensive	little
$ZnCl_2$	0.18	548	1029	some	little
$HgCl_2$	0.17	277	304	some	little
HCl	0.16	−114	−85	none	none
BCl_3	0.14	−107	12	complete	little
$InCl_3$	0.14	586	s	extensive	little
$SiCl_4$	0.13	−68	57	complete	little
ICl	0.13	27	–	extensive	some
$TlCl_3$	0.12	d	–	extensive	some
$GaCl_3$	0.11	78	200	extensive	some
$BiCl_3$	0.11	232	441	extensive	some
$SbCl_3$	0.10	73	221	extensive	some
PCl_3	0.10	−92	76	complete	some
$SnCl_4$	0.10	−33	113	extensive	some

TABLE 9-1 (continued)

Compound	$-\delta_{Cl}$	mp, °C	bp, °C	Hydrolysis	Oxidizing or chlorinating ability
PbCl$_4$	0.10	−15	d	extensive	high
TeCl$_4$	0.07	224	338	extensive	high
GeCl$_4$	0.07	−49	83	extensive	some
PCl$_5$	0.07	160(P)	−	complete	high
SbCl$_5$	0.07	73	−	extensive	high
CCl$_4$	0.06	−23	77	complete	high
AsCl$_3$	0.06	−16	130	extensive	some
BrCl	0.04	−	−	complete	high

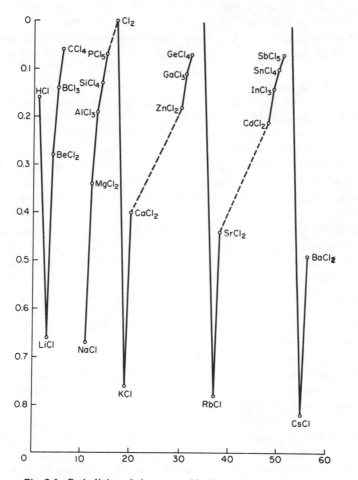

Fig. 9-1 Periodicity of charge on chlorine in binary chlorides.

BONDS AND BOND ENERGIES IN HALIDES

Gaseous Halides

Data relevant to the calculation of atomization energies of gaseous halides are presented in Tables 9-2, 9-3, 9-4, and 9-5 for fluorides, chlorides, bromides, and iodides.

Unstable Monohalides.   Let us first consider the bonding in those molecules which lack stable existence under ordinary conditions, such as the monohalides of polyvalent elements. Unfortunately, the experimental data are not very reliable for these gas molecules, so the difference between experimental and calculated atomization energies is of uncertain meaning. Furthermore, important data are lacking. However, we may conclude, very tentatively, that the M2 monohalide molecules can be described as held together by ordinary single polar covalent bonds. There is no consistent evidence of any promotional energy requirement to unpair the s electrons other than may be automatically compensated by the evaluation of electronegativity.

In the M3 monohalides, there should be no need of promoting an s electron from the ground state structure, s^2, p, for only the p electron is needed. Then the atomization energy should be somewhat higher than calculated. For AlF the difference is about 25 kpm, but the difference is negligible for BF. For the chlorides and bromides, it appears necessary to invoke double bonds from Al, Ga, and In to halogen in order to account for the experimental bond energy. However, it is also possible that the experimental values are higher than those calculated for a normal single bond because there is no need for promotional energy in forming the monohalide.

The unsatisfactory nature of the data is also illustrated by the silicon halides. An atom of silicon, with ground state of the outermost level indicated to be s^2, p_x, p_y, should be able to form monohalide and dihalide without any promotional energy but should require it for the trihalide and tetrahalide. The compounds, SiF and SiF_2, are reasonably well described as having Si-F''' bonds, but the experimental values for SiF_3 and SiF_4 appear significantly higher than the calculated values. For the chlorides, SiCl and $SiCl_4$ are well described as having Si-Cl''' bonds, but the calculated energy for the $SiCl_2$ molecule is too low.

It appears reasonable that, where the important parameters of an atom have been determined on the basis of its "normal" valences, differences might be observed for lower states requiring less promotional energy. However, one must conclude from the available data that the evidence is inconsistent.

The common occurrence of removal of the lone pair weakening effect on the halogen is notable. It appears to reflect the presence of outer orbital vacancies on the other atom which somehow allow the halogen atom to form the strongest single bond of which it is capable. This has previously been ascribed to "π bonding" in which the bond is supplemented by lone pair electrons of the halogen which are

TABLE 9-2
Atomization of Gaseous Fluorides

Compound	$-\delta_F$	R_0	E_{calc}	E_{exp}	Bond type
HF	0.25	0.92	135.5	135.8	H–F'
LiF	0.74	1.56	168.7	137	Li–F'
NaF	0.75	1.93	136.7	115	Na–F'
KF	0.84	2.17	132.9	121, 123	K–F'
RbF	0.86	2.27	129.7	121, 127	Rb–F'
CsF	0.90	2.35	129.7	123, 131	Cs–F'
BeF	0.47	1.36	142.6	137	Be–F'
BeF$_2$	0.34	1.43	308.0	303.6	Be–F"
MgF	0.55	1.75	123.9	111 ± 5	Mg–F'
MgF$_2$	0.41	1.77	262.4	258.9	Mg–F'
			275.2		Mg–F"
CaF	0.62	1.93	124.2	127	Ca–F'
CaF$_2$	0.47	2.02	266.6	268.3	Ca–F"
SrF	0.66	2.08	120.5	130	Sr–F'
			130.8		Sr–F'''
SrF$_2$	0.50	2.10	69.8	263	Sr–F"
BaF	0.73	2.16	122.6	141	Ba–F'
			129.7		Ba–F'''
BaF$_2$	0.57	2.68	274.0	273	Ba–F"
ZnF$_2$	0.23	1.81	184.6	183.8	Zn–F'
BF	0.33	1.26	128.4	132	B–F'
BF$_2$	0.23	1.30	306.6	302.3	B–F'''
BF$_3$	0.18	1.30	469.3	463.0	B–F'''
AlF	0.44	1.64	135.1	160, 158	Al–F'''
AlF$_2$	0.31	1.64	279.1	272.8	Al–F'''
AlF$_3$	0.24	1.63	428.4	421.2	Al–F'''
CF	0.22	1.27	108.5	115–128	C–F'
			128.1		C–F"
CF$_2$	0.15	1.30	253.7	250	C–F"
CF$_3$	0.12	1.32	356.1	344.5	C–F', 2C–F"
CF$_4$	0.09	1.32	462.4	465	2C–F', 2C–F"
SiF	0.34	1.60	131.0	129	Si–F'''
SiF$_2$	0.24	1.49	282.9	289	Si–F'''
SiF$_3$	0.12	1.54	414.5	433.1	Si–F'''
SiF$_4$	0.10	1.54	557.2	570.5	Si–F'''
GeF	0.24	1.68	111.0	118	Ge–F'''
GeF$_4$	0.10	1.68	453.1	450.0	Ge–F'''
SnF	0.55	1.94	112.6	91	Sn–F'
PbF	0.41	2.06	83.6	85	Pb–F"
PbF$_2$	0.29	2.11	183.3	188.6	Pb–F"
PbF$_4$	0.14	2.08	309.5	308.4	2Pb–F', 2Pb–F"
NF	0.14	1.37	68.9	73.3	N'–F'
NF$_2$	0.08	1.37	137.8	140.7	N'–F'
NF$_3$	0.07	1.37	206.7	200	N'–F'
N$_2$F$_4$	0.08	1.37	311.8	303.3	N'–F', N'–N'
PF	0.26	1.78	116.8	112.3	P–F'''
PF$_2$	0.18	1.55	242.1	222.2	P–F'''

TABLE 9-2 (continued)

Compound	$-\delta_F$	R_0	E_{calc}	E_{exp}	Bond type
PF_3	0.14	1.54	369.2	357.9	P—F'''
PF_5	0.09	1.57	553.9	555.1	3P—F''', 2P—F'
AsF_3	0.11	1.71	302.4	349.0	As—F'''
AsF_5	0.08	1.71	468.7	462.4	2As—F', 3As—F'''
BiF	0.30	2.05	73.9	60 ± 5	Bi—F'
OF	0.05	1.42	45.3	46.1	O'—F'
OF_2	0.03	1.42	90.6	89.8	O'—F'
S_2F_2	0.18	1.60	203.1	–	S—F'
SF_2	0.12	1.59	157.8	–	S—F'
SF_4	0.08	1.56	328.1	327.4	S—F'
SF_6	0.05	1.56	492.0	471.8	S—F'
SeF_6	0.05	1.70	424.2	430	Se—F'
TeF_6	0.07	1.84	461.4	474	Te—F'
ClF	0.09	1.63	63.1	59.9	Cl'—F'
ClF_3	0.04	1.60, 1.70	124.8	124.7	Cl'—F', 2 half-bonds
ClF_5	0.03	1.60, 1.72	185.3	181.4	Cl'—F', 4 half-bonds
BrF	0.13	1.76	62.9	59.6	Br'—F'
BrF_3	0.07	1.81	140.6	144.5	Br—F'', 2 half-bonds, F'
BrF_5	0.04	1.68, 1.78	217.8	223.7	Br—F''', 4 half-bonds, F'
IF	0.21	1.91	67.9	67.0	I'—F'
IF_5	0.07	1.75, 1.86	321.1	320.0	3I'—F'', 2 half-bonds
IF_7	0.06	1.75, 1.86	329.3	386.9	3I'—F'', 4 half-bonds

shared with the other atom. The evidence from bond energy calculations, in which the weakening effect affects only the nonpolar covalent contribution, consistently suggests that the increased bond strength does not arise from an increase in the number of shared electrons, but rather only in a decrease of the bond weakening effect. Among the four tables are 56 different compounds in which the bonding appears, according to calculations, to be $-X'''$.

"Normal" Halides.   For some undetermined reason, the calculated bond energies for LiF, NaF, and KF, LiCl, NaCl, LiBr, NaBr, LiI, and NaI gas molecules appear significantly too high. The calculated values for the other eleven alkali halide gas molecules are in good agreement with the experimental values. There is no obvious cause of this discrepancy. It would seem quite unreasonable to ascribe it to the method of calculation, which works so well otherwise. It will be noted that no such discrepancy appears in the data for the solid halides, wherein reasonable agreement between calculated and experimental atomization energies is observed without exception.

The boron halides provide the only exception to the rule that wherever individual molecules possess both outer vacant orbitals and lone pair electrons, further condensation to dimers or higher aggregates occurs. The reason for this exception is not known. One may speculate that steric requirements of a halogen bridge in boron halides analogous to aluminum halides might cause the monomer to be more stable. This might be true especially if the condensation should reinstate the lone pair weakening of the halogen by destroying the vacancy which removed that weakening in the monomer.

The data for carbon fluorides are especially interesting, as they appear to provide the only example of reduction in lone pair weakening effect where there are no low energy vacancies on the other atom (the carbon). The experimental bond energies suggest that the first two carbon to fluorine bonds are $C-F''$. The subsequent two bonds appear to be $C-F'$, averaged, of course, with the first two. Perhaps the electron removal from the carbon atom is here sufficient to activate orbitals of the $n = 3$ shell so that they can cooperate in reducing the lone pair weakening effect somewhat.

The atomization energy of carbon tetrachloride is found experimentally to be significantly lower (by about 36 kpm) than the calculated value. The internuclear distance between chlorine atoms in CCl_4 is considerably less than the so-called van der Waals radius of the chlorine would permit. Presumably this crowding is what weakens the bonding. This explanation is consistent with the fact that CS_2 does not polymerize. In contrast to CO_2, the bonding in CS_2 polymer would indeed be stronger than in the equivalent amount of monomer. The reason for CS_2 not polymerizing is therefore different from the reason for CO_2 not polymerizing. It appears to be that crowding four sulfur atoms around each carbon, sulfur atoms being about the same size as chlorine atoms, would weaken the bonding to the extent of favoring the monomer. Calculations bear this out.

The capability of CCl_4 to act as chlorinating agent, especially at higher temperatures, is very consistent with the low charge on the combined chlorine, only -0.06. The need for higher temperature arises from the kinetic difficulty of attacking a CCl_4 molecule. One can calculate that hydrolysis should occur exothermically and to completion, but we are all familiar with the inertness of CCl_4 to water under ordinary conditions. Here is an example of kinetic rather than thermodynamic stability.

It may be noted that $SiBr_4$, wherein the relative atomic sizes may be likened to CCl_4, also has a lower experimental atomization energy than calculated, although the values for $SiCl_4$ are in good agreement. Presumably the atomization energy of $SiBr_4$ is also reduced somewhat by crowding of the halogen atoms.

The conversion of phosphorus trihalides to pentahalides has usually been pictured as involving promotion of one of the lone pair electrons of the phosphorus to an outer d orbital, followed by hybridization of some nature and the formation of two additional covalent bonds of normal type. We find for PF_3 and PF_5 the suggestion that the two extra bonds are $P-F'$ bonds, compared with the first three bonds which are $P-F'''$. The situation seems to be similar for the arsenic fluorides, although the experimental value for AsF_3 seems far too high. However, in the phosphorus chlorides the situation appears different, presumably involving two half-bonds as the final two bonds in PCl_5. The data are not conclusive.

A calculation of the bond energy in SF_2 and related compounds shows why this is not a well-known compound. SF_2 has been prepared in a transitory state, but

TABLE 9-3
Atomization of Gaseous Chlorides

Compound	$-\delta_{Cl}$	R_0	E_{calc}	E_{exp}	Bond type
HCl	0.16	1.27	103.7	103.3	H—Cl′
LiCl	0.65	2.03	122.6	114.3	Li—Cl′
NaCl	0.67	2.36	105.8	99.0	Na—Cl′
KCl	0.76	2.67	101.9	101.6	K—Cl′
RbCl	0.78	2.79	99.5	100.5	Rb—Cl′
CsCl	0.81	2.91	97.9	101.	Cs—Cl′
BeCl	0.39	1.75	110.3	93.	Be—Cl′
BeCl$_2$	0.28	1.77	225.8	222.8	Be—Cl′
MgCl	0.47	2.18	97.4	–	Mg—Cl′
MgCl$_2$	0.34	2.18	203.2	204.3	Mg—Cl′
CaCl	0.54	2.44	96.5	97	Ca—Cl′
ZnCl$_2$	0.16	2.09	159.8	157.2	Zn—Cl′
BCl	0.24	1.72	109.1	118 ± 9	B—Cl‴
BCl$_2$	0.17	1.73	211.0	213.4	B—Cl‴
BCl$_3$	0.13	1.75	316.2	318.3	B—Cl‴
B$_2$Cl$_4$	0.17	1.73	464.3	502.6	B—Cl‴
AlCl	0.35	2.13	117.3	118	avg. Al—Cl′, Al=Cl″
AlCl			95.5		Al—Cl‴
AlCl$_3$	0.19	2.06	307.5	305.0	Al—Cl‴
GaCl	0.20	2.20	114.4	114	Ga=Cl″
GaCl$_3$	0.10	2.09	254.1	260.5	Ga—Cl‴
InCl	0.26	2.32	108.8	102.4	In=Cl″
InCl$_3$	0.14	2.46	222.0	221.3	In—Cl‴
CCl	0.13	1.65	85.1	68.4	C—Cl′
CCl$_4$	0.06	1.77	347.6	312.3	C—Cl′
SiCl	0.26	2.00	95.0	92.6	Si—Cl‴
		2.06	92.2		
SiCl$_2$	0.18	2.00	191.8	206.7	Si—Cl‴
SiCl$_4$	0.11	2.02	380.0	382.3	Si—Cl‴
GeCl	0.16	2.08	81.1	82.1	Ge—Cl‴
GeCl$_4$	0.07	2.09	322.4	324.9	Ge—Cl‴
SnCl$_4$	0.10	2.31	308.0	307.1	Sn—Cl‴
PbCl	0.32	2.46	70.5	72.4	Pb—Cl‴
PbCl$_4$	0.09	2.43	237.6	238.2	Pb—Cl′
NCl$_3$	0.02	1.73	165.0	–	N—Cl′
PCl	0.18	2.04	75.1	73.5	P—Cl′
PCl$_3$	0.09	2.04	225.3	231.1	P—Cl′
PCl$_5$	0.06	2.19	311.0	310.3	3P—Cl″, 2 half-bonds
		2.04			
AsCl$_3$	0.07	2.16	214.2	221.4	As—Cl‴
SbCl$_3$	0.10	2.33	208.2	225.0	Sb—Cl‴
BiCl$_3$	0.11	2.48	201.3	200.0	Bi—Cl‴
S$_2$Cl$_2$	0.09	1.99	185.8	195.8	S—Cl′
SCl$_2$	0.06	2.00	134.4	129.5	S—Cl′
SCl$_4$	0.04	2.00	268.8	–	S—Cl′
SeCl$_2$	0.05	2.15	117.8	115.0	Se—Cl′
BrCl	0.04	2.14	55.6	51.3	Br—Cl′
ICl	0.12	2.30	58.0	50.3	I—Cl′

TABLE 9-4
Atomization of Gaseous Bromides

Compound	$-\delta_{Br}$	R_0	E_{calc}	E_{exp}	Bond type
HBr	0.12	1.41	87.6	87.5	H−Br′
LiBr	0.61	2.17	109.4	101.9	Li−Br′
NaBr	0.62	2.50	94.0	88.8	Na−Br′
KBr	0.71	2.82	91.5	91.0	K−Br′
RbBr	0.73	2.95	89.4	90.4	Rb−Br′
CsBr	0.77	3.07	89.1	91	Cs−Br′
BeBr	0.35	1.84	98.5	122	Be−Br′
BeBr$_2$	0.25	1.90	202.0	190.6	Be−Br′
MgBr	0.42	2.36	83.8	82	Mg−Br′
MgBr$_2$	0.31	2.34	177.2	173.4	Mg−Br′
ZnBr$_2$	0.13	2.24	137.4	131.8	Zn−Br′
BBr	0.20	1.87	88.3	99.6	B−Br‴
BBr$_2$	0.14	1.87	179.0	176.8	B−Br‴
BBr$_3$	0.10	1.87	268.5	264.3	B−Br‴
AlBr	0.31	2.30	81.5	100−108	Al−Br‴′
			119.3		Al=Br″
AlBr$_3$	0.17	2.27	253.5	256.4	Al−Br‴
GaBr	0.15	2.35	95.7	99.3	Ga=Br″
InBr	0.21	2.54	92.3	93	In−Br‴
InBr$_3$	0.11	2.58	192.6	192.7	Si−Br‴
SiBr$_4$	0.09	2.15	326.0	315.0	Ge−Br‴
GeBr	0.11	2.30	64.7	60.4	Ge−Br‴
GeBr$_4$	0.05	2.30	258.8	268.5	Sn−Br‴
SnBr$_4$	0.08	2.44	260.4	254.2	Pb−Br‴
PbBr	0.28	2.60	60.6	60.2	Pb−Br‴
PbBr$_2$	0.20	2.60	125.0	124.4	Pb−Br‴′
PbBr$_4$	0.08	2.58	198.8	197.6	Pb−Br‴
PBr	0.13	2.23	61.9	63.1	P−Br′
PBr$_3$	0.07	2.20	190.8	188.7	P−Br′
AsBr$_3$	0.04	2.33	175.8	183.5	As−Br‴
BiBr$_3$	0.09	2.63	167.4	168.1	Bi−Br‴
IBr	0.08	2.45	48.6	42.4	I′−Br′

it readily changes to sulfur and higher fluorides. This can easily be shown to be a favored reaction because the S−F bonds are not changed in number nor much in energy, while sulfur with its stable bonds is being liberated, thus providing greater overall stability. Similarly, it can be shown that sulfur iodides do not exist because, in this rather rare example, the bonds are actually stronger in the free elements than in the compound.

In connection with the general problem of whether promotional energies are significant, hybridizations important, and valence changes influential, it is especially interesting to note that the same parameters for sulfur appear applicable to H_2S, SO_2, H_2SO_4, SF_4, and SF_6, among others. The calculated atomization energies of SeF_6 and TeF_6 are a little lower than the experimental values, whereas the experimental value is lower in SF_6. This suggests the possibility that there may be a

TABLE 9-5

Atomization of Gaseous Iodides

Compound	$-\delta_{Br}$	R_0	E_{calc}	E_{exp}	Bond type
HI	0.04	1.61	70.4	71.3	H–I″
LiI	0.53	2.39	90.4	85.7	Li–I′
NaI	0.54	2.71	78.6	72.2	Na–I′
KI	0.63	3.05	77.3	77.8	K–I′
RbI	0.65	3.18	76.0	76.8	Rb–I′
CsI	0.69	3.32	75.8	75	Cs–I′
BeI	0.26	2.05	80.6	102.8	Be—I‴
BeI_2	0.19	2.12	155.6	152.3	Be–I′
ZnI_2	0.07	2.42	106.8	100	Zn–I′
BI	0.12	2.10	63.8	81.7	B–I′
BI_2	0.08	2.10	127.6	131.0	B–I′
BI_3	0.06	2.03	197.7	194.1	B–I′
AlI_3	0.12	2.44	205.8	203.5	Al—I‴
InI_3	0.07	2.80	149.4	150.5	In—I‴
GeI_4	0.01	2.49	200.8	205.6	Ge—I‴
PbI	0.20	2.79	47.9	46.6	Pb—I‴
PbI_2	0.14	2.79	97.6	97.6	Pb—I‴
PbI_4	0.04	2.77	147.2	149.2	Pb–I′
PI_3	0.03	2.47	140.7	–	P–I′
AsI_3	0.00	2.52	130.8	–	As–I′
BiI_3	0.04	2.84	128.7	129.9	Bi—I‴
SI_2	0.02	2.37	95.0	–	S–I′

little crowding, absent in the larger atoms, of the six fluorine atoms around the sulfur.

Finally, we come to the interhalogen compounds. These have long been thought of as involving promotion of lone pair electrons to outer d orbitals, wherever the valence is increased beyond one. The bond energy data suggest otherwise. They do not exhibit an overall consistency, as can be seen from the interpretations provided in the tables. However, they do seem to require an interpretation involving some half-bonds. This would account very nicely for the very high halogenating power of some of the polyhalides. For example, it only requires about 33 kpm of fluorine atoms to liberate them from ClF_3. This is more than the 19 required to form atoms from F_2 molecules but still quite low.

The greatest importance of these compounds, however, may be their exemplification of three-center bonding of the type postulated for the xenon fluorides and perhaps to be found also in the bifluoride ion and in PCl_5, or perhaps even SF_6.

From the thermal data of these compounds it is possible to explain the tendency for the lower interhalogens to disproportionate to form the higher polyhalides such that, for example, the higher fluorides of bromine and iodine are much more stable than the lower ones.

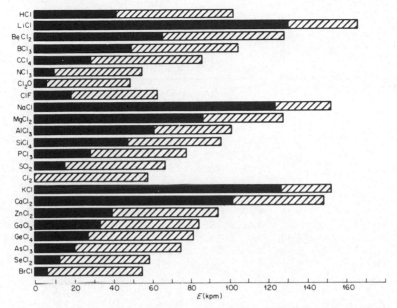

Fig. 9-2 Periodicity of chloride bond energies. Black areas represent ionic contribution.

Solid Halides

Data relevant to the bond energy calculations for solid halides are presented in Tables 9-6, 9-7, 9-8, and 9-9. These calculations are all based on the coordinated polymeric model and demonstrate how effective this model is over a considerable range of compounds.

TABLE 9-6
Atomization of Solid Fluorides

Compound	n	$-\delta_F$	M	k	R_0	$t_c E_c$	E_{calc}	E_{exp}
LiF	3	0.74	1.75	0.85	2.01	26.2	208.0	203.8
NaF	3	0.75	1.75	0.87	2.31	19.2	183.3	181.8
KF	4	0.84	1.75	0.88	2.66	14.5	176.0	175.8
RbF	4	0.86	1.75	0.89	2.82	12.5	170.2	169.8
CsF	4	0.90	1.75	0.90	3.01	8.3	164.6	164.5
BeF_2	3	0.34	4.44	0.84	1.61	88.2	349.7	359.4
MgF_2	4	0.41	4.76	0.84	2.02	89.1	358.5	352.0
CaF_2	4	0.47	5.04	0.87	2.36	80.0	369.9	370.6
SrF_2	4	0.50	5.04	0.87	2.51	74.1	364.1	367.1
BaF_2	4	0.57	5.04	0.88	2.68	53.4	366.6	366.7
ZnF_2	3	0.23	4.82	0.84	2.04	92.8	244.4	244.8
CdF_2	3	0.27	5.04	0.87	2.32	72.3	241.7	229.8
PbF_2	4	0.29	4.62	0.86	2.29	76.0	243.0	244.2

TABLE 9-7

Atomization of Solid Chlorides

Compound	n	$-\delta_{Cl}$	M	k	R_0	$t_c E_c$	E_{calc}	E_{exp}
LiCl	3	0.65	1.75	0.88	2.57	38.6	167.9	165.2
NaCl	3	0.67	1.75	0.89	2.81	28.8	152.2	153.2
KCl	4	0.76	1.75	0.90	3.14	24.9	151.6	154.6
RbCl	4	0.78	1.75	0.91	3.28	22.7	148.4	151.6
CsCl	6?	0.81	1.76	0.90	3.45	27.6	151.1	151.3
$BeCl_2$	3	0.28	4.09	0.85	2.02	111.3	271.3	258.9
$MgCl_2$	3	0.34	4.49	0.87	2.54	83.1	256.7	256.9
$CaCl_2$	3	0.40	4.73	0.89	2.74	80.0	284.0	290.7
$SrCl_2$	4	0.43	4.62	0.90	3.02	95.5	292.1	295.2
$BaCl_2$	4	0.49	5.0	0.90	3.18	72.6	302.8	305.5
$ZnCl_2$	3	0.16	4.49	0.87	2.64	109.7	188.3	188.6
$CdCl_2$	3	0.21	4.49	0.89	(2.79)	–	–	–
$AlCl_3$	3	0.19	8.30	0.87	2.16	129.6	340.4	333.6
TlCl	4	0.40	1.76	0.91	3.31	57.7	122.0	121.7
$PbCl_2$	4	0.23	4.62	0.88	2.98avg.	87.7	191.9	191.2

TABLE 9-8

Atomization of Solid Bromides

Compound	n	$-\delta_{Br}$	M	k	R_0	$t_c E_c$	E_{calc}	E_{exp}
LiBr	3	0.61	1.75	0.88	2.75	38.1	141.5	149.0
NaBr	3	0.62	1.75	0.90	2.98	29.5	138.3	138.6
KBr	4	0.71	1.75	0.91	3.29	26.8	141.0	138.6
RbBr	4	0.73	1.75	0.91	3.43	24.8	137.3	139.3
CsBr	6?	0.77	1.76	0.91	3.71	28.8	139.2	139.7
	4	–	–	–	3.60	19.8	132.9	–
$BeBr_2$	3	0.13	4.09	0.85	2.16	122.1	229.0	220.1
$MgBr_2$	3	0.21	4.38	0.88	2.70	77.3	224.2	222.9
$CaBr_2$	3.5?	0.24	4.73	0.89	2.94	87.1	258.2	257.3
$SrBr_2$	4	0.26	4.62	0.90	3.21	90.0	257.7	263.6
$BaBr_2$	4	0.30	5.0	0.91	3.38	68.8	269.9	275.8
$ZnBr_2$	3	0.08	4.49	0.88	2.75	103.7	165.7	162.6
$PbBr_2$	4	0.13	4.62	0.89	3.19	80.6	166.2	166.3

TABLE 9-9
Atomization of Solid Iodides

Compound	n	$-\delta_I$	M	k	R_0	$t_c E_c$	E_{calc}	E_{exp}
LiI	3	0.53	1.75	0.90	3.03	40.2	131.3	128.5
NaI	3	0.54	1.75	0.90	3.23	31.2	118.8	120.2
KI	4	0.63	1.75	0.91	3.53	31.0	125.4	125.1
RbI	4	0.65	1.75	0.92	3.66	28.1	123.0	123.6
CsI	4	0.69	1.76	0.92	3.82	23.4	120.5	124.7
BeI_2	3	0.19	4.09	0.87	2.37	99.2	193.9	179.9
MgI_2	3	0.24	4.38	0.89	2.94	74.4	180.0	180.0
CaI_2	4	0.30	4.38	0.90	3.04	99.5	228.7	221.3
SrI_2	4	0.33	4.62	0.91	3.42	87.1	221.8	225.5
BaI_2	4	0.39	5.0	0.91	3.59	67.5	231.6	237.0
ZnI_2	3	0.05	4.38	0.89	2.95	99.6	130.3	132.0
PbI_2	4	0.09	4.38	0.90	3.13	83.7	142.3	139.7

ATOMIZATION OF OXYHALIDES

Table 9-10 lists data for some oxyhalide atomization energies. They show that coordination of oxygen as acceptor seems to result in reduction of the lone pair weakening effect on oxygen, in combinations such as these wherein the other element has outer d orbital vacancies. A study of these data emphasizes the necessity of further study of the so-called weakening effect.

TABLE 9-10
Atomization of Oxyhalides

Compound	Bond	δ_X	R_0	E_{calc}	E_{exp}
POF_3	P–F'''	−0.13	1.52	483.8	481.0
	P–O''	−0.03	1.45		
$POCl_3$	P–Cl'	−0.07	1.99	355.3	355.6
	P–O'''	−0.12	1.45		
$POBr_3$	P–Br'	−0.03	2.16	314.3	312.2
	P–O'''	−0.17	1.45		
SOF_2	S–F'	−0.12	1.53	276.0	277.0
	S–O'''	−0.01	1.41		
$SOCl_2$	S–Cl'	−0.03	2.07	239.3	235.2
	S–O'''	−0.09	1.45		
SO_2Cl_2	S–Cl'	−0.02	1.99	330.0	331.0
	S–O''	−0.07	1.43		
$SOBr_2$	S–Br'	0.01	2.18	203.6	201.4
	S–O''	−0.13	1.45		
$PSCl_3$	P–Cl'	−0.11	2.02	318.0	320.1
	P–S'''	0.07	1.85		

TEN

Bond Dissociation

THE INTERRELATIONSHIP BETWEEN CONTRIBUTING
BOND ENERGY AND BOND DISSOCIATION ENERGY

The bond energies that have been the focus of major attention so far in this book are perhaps best called "contributing bond energies" (CBE). This is because each energy represents the contribution made by that bond to the total atomization energy of the molecule. Presumably, if the molecule could be atomized instantaneously breaking completely into separate atoms at one simultaneous explosion, then the energy required to break each bond would be its CBE.

This event can be achieved only for diatomic molecules, since these have but one bond to break. When the number of bonds in a molecule exceeds one, then they tend to be broken one at a time. A source of considerable confusion for many years has been the recognition that the energy to break one bond of a polybonded aggregate is seldom that expected, based on the presumed (CBE) strength of the bond. For example, the atomization of water is known to require 221.6 kpm. There being no basis for suspecting any difference between the two O–H bonds, the energy of each is evaluated as half the total, or 110.8 kpm. This is the CBE of O–H in water. Yet careful measurement shows that it takes 119.2 kpm to break the first bond and only 102.4 kpm to break the second. These are called the "bond dissociation energies" (BDE). Their sum, of course, is the same as the sum of the two CBEs, in keeping with the conservation of energy. But it is clear that it takes 8.4 kpm more than 110.8 kpm to break the first bond, because this process somehow weakens the second or remaining O–H bond by that amount, $110.8 - 8.4 = 102.4$ kpm. In other words, breaking the first bond is more difficult because it simultaneously weakens the second bond, requiring enough energy to do both. Before examining in more detail why this should be, let us be aware that this is a very common phenomenon—that breaking one bond usually affects the residual bond strength of a fragment in one way or another, thus contributing to the energy of breaking that bond exothermically or endothermically.

During the writing of the first edition, I was of course aware of these mysterious fluctuations in BDE but postponed the study of bond dissociation in order to concentrate more fully on the narrower, but still amply formidable, problem. As soon as possible, I then directed attention to reconciliation of the two seemingly separate areas of information. For it is one thing to understand the polar nature of a C—Cl bond, for example, so that one can calculate with quantitative accuracy its contribution to the total atomization energy of the compound. But when one has calculated the energies of two chlorine to carbon bonds, each in a different environment, and found them to be very nearly equal, then it is disconcerting to realize that the bond dissociation energy, required to break loose a chlorine atom from such a bond, may differ very substantially between the two bonds. For instance, the C—Cl bonds in CH_3Cl and $C_6H_5CH_2Cl$ have like CBEs of 83 kpm. Yet the BDE is 85 for the first compound and only 70 for the second. Why is this so?

It is reasonable to assume that the total atomization energy of a molecule must be independent of the manner in which the molecule may be taken apart, as long as individual atoms are the final product. It has been determined that the sum of the contributing bond energies equals the total atomization energy. It is equally true that the sum of the energies required for the stepwise atomization of the compound must equal the same total atomization energy. It follows that if breaking one bond requires different energy from that calculated as the CBE, this difference must be accounted for in the energy content of the fragments formed by the bond rupture. For example, breaking the first bond in water costs about 8 kpm more than expected on the basis of the CBE, or the average bond energy. From this fact, we may conclude that breaking this first bond must in a sense also contribute 8 kpm toward the breaking of the remaining bond. This is verified when the bond dissociation energy for the second bond is found to be 8 kpm less than the original CBE.

The interrelationship between bond dissociation energy and contributing bond energy can be expressed simply:[1]

$$BDE = CBE + E_R (I) + E_R (II)$$

E_R stands for "reorganizational energy." It is the energy involved in the reorganization of the electronic structure, and possibly the geometric structure as well, of a free radical when it is created by homolytic release from a covalent bond. If either fragment (I) or fragment (II) is a **single atom, its reorganizational energy is zero.** But a fragment containing more than one atom will undergo a further change as the newly released former valence electron interacts in the free radical system. The reorganizational energy will be either positive or negative, depending on the

[1] R. T. Sanderson, *J. Am. Chem. Soc.* 97, 1367 (1975).

circumstances. Wherever it is possible for the free radical to rearrange into a stable molecule, E_R will be negative and easily determinable quantitatively from a knowledge of the thermochemistry of the stable molecule. In other words the bond dissociation will be favored, requiring less energy, by the amount of energy released in conversion to the molecule. However, where formation of a stable molecule is not possible, the electronic changes and perhaps the steric changes that may accompany them are more subtle and more difficult to predict or evaluate. They may be either exothermic or endothermic.

In the example cited above, the reorganizational energy is zero for liberation of a chlorine atom but +2 for a methyl radical and −13 for a benzyl radical. Thus the breaking loose of a chlorine atom from methyl chloride takes $83 + 2 + 0 = 85$ kpm, whereas the similar dissociation from benzyl chloride takes $83 − 13 + 0 = 70$ kpm. But what is the source of these reorganizational energies?

REORGANIZATIONAL ENERGIES OF RADICALS

In a spirit of speculation and adventure, let us imagine how things might be and then compare with how they are. A more logical and prudent approach would seem to be the reverse, in which one is well advised to find out how things are before he tries to explain them. However, as will be seen shortly, the experimental evidence is tantalizingly imprecise and the possibility of progress should justify a certain risk.

Let us assume that the dissociation of a bond in a molecule having more than one bond involves two processes:

(1) Transformation from compound molecule to two radicals, each with an outermost half-filled orbital that has been released from its bonding function.

(2) Reorganization of the half-filled orbital within each radical system so that maximum radical stability is achieved.

Let us further assume that process (1) is completely and perfectly covered by the CBE of that bond. Whatever polarity might have been present in the bond has been withdrawn to create perfect neutrality in each of the two fragments. For a diatomic molecule, this CBE is of course identical with the BDE. Process (2) begins, however, where process (1) ends. The liberated bonding orbital, with its one valence electron, may become an asset to the radical, entering into its electronic system in such a way as to reinforce the residual bonding. This will result in release of reorganizational energy. On the other hand, there may be no stable accommodation for this released half-filled orbital, which can only interfere with the rest of the radical now that it is no longer constrained by its own bonding function. Such activity would destabilize the radical such that the reorganizational energy would be positive and absorbed. Possible mechanisms for this kind of destabilization will be considered presently.

One may assume that any given radical, once liberated from whatever combination, to the extent indicated by the magnitude of the contributing bond energy, would be in a condition determined solely by its composition and structure and entirely independent of the nature of its previous very recent association. There seems little question that in the final, most stable state of the radical, it should also be independent of any factors other than those related to its composition and structure. The difference in energy content between the two conditions of the radical must therefore also be constant, characteristic of the radical alone and independent of bonding conditions that preceded the formation of the radical. If this is so, evaluation of "constant" reorganizational energies should be possible for each radical such that the bond dissociation energy for any bond that involves this radical can be accurately determined by adding the reorganizational energies of the two radicals to the contributing bond energy.

To illustrate further by the preceding example of the methyl and benzyl chlorides, if liberation of the methyl radical increases the bond dissociation energy by 2 kpm for methyl chloride, it will increase the BDE similarly for every methyl compound. Likewise, if liberation of the benzyl radical reduces the requirement of BDE by 13 kpm in the chloride, it will contribute similarly for all other benzyl compounds. Then what about combining benzyl and methyl? This would create a new C—C bond, for which the CBE would be about 83 kpm. But dissociation of this bond would only require 72 kpm: BDE = 83 + 2 − 13 = 72 kpm.

As hinted earlier, the experimental determination of bond dissociation energies is difficult and the results are not often highly accurate. It is possible to obtain, however, a reasonably self-consistent set of bond dissociation energies from heats of formation of both compounds and radicals using the "best" values of the literature.[2,3] Combinations of these can be augmented by compounds of the radicals with hydrogen or halogen, principally, as single atoms. I have been able to collect data for about 200 compounds, upon which the study summarized in Table 10-1 is based. The original study concluded that the available data were too uncertain to serve as a basis for deciding whether reorganizational energies of radicals are indeed constant. For example, the values that appeared applicable to methyl radical varied from about −5 to +3 kpm. Yet a study of these variations revealed no consistent trends but quite a random nature. The possibility of constant reorganizational energies was not ruled out by the literature data. Neither was it

[2] Compounds: J. D. Cox and G. Pilcher, "Thermochemistry of Organic and Organometallic Compounds," Academic Press, New York, 1970; D. R. Stull, E. F. Westrum Jr., and G. C. Sinke, "The Chemical Thermodynamics of Organic Compounds," Wiley, New York, 1969.

[3] Radicals and BDE work: S. W. Benson, *J. Chem. Ed.* **42**, 502 (1965); J. A. Kerr, *Chem. Rev.* **66**, 465 (1966); D. D. Wagman, W. H. Evans, V. B. Parker, I. Halow, S. M. Bailey, and R. H. Schumm, "Selected Values of Chemical Thermodynamic Properties," NBS Technical Note 270–3, National Bureau of Standards, 1968; B. D. Darwent, "Bond Dissociation Energies in Simple Molecules," NSRDS—NBS 31, National Bureau of Standards, 1970; J. A. Kerr and A. F. Trotman—Dickenson, "CRC Handbook of Chemistry and Physics," **53**, F189 (1974).

clearly indicated. However, by a careful procedure of judicious selection and appropriate averaging, it was found possible to assign energy of reorganization to each radical such that the experimental bond dissociation energy can be quite accurately represented as the sum of the calculated CBE and the standard E_R values.

WHAT IS THE EVIDENCE?

The evaluation of experimental bond dissociation energy as the difference between the heats of formation of the compound and those of the radicals depends on values of heats of formation of the compound sometimes known only to about 1 kpm or more and those of the radicals perhaps averaging 2 kpm uncertainty. The evaluation of CBE depends on all the factors involved in the quantitative calculation of polar covalent energy. It was assumed that if the total calculated atomization energy of a compound is in good agreement with the experimental value, each individual calculated bond energy is probably correct. However, infrequently the calculated atomization energy differs by a few kpm from the experimental value. In such cases, it was assumed that the experimental value was correct, and the individual bond energy of interest was corrected accordingly. This made it at least consistent with the experimental heat of formation used in determining the bond dissociation energy.

Data for several different combinations of each radical were then used to evaluate an average reorganizational energy, which after further refinement became the tabulated value. The average number of examples of each radical in Table 10-1 is 12. Table 10-1 also lists, after the formula of each radical at the head of each group, the standard heat of formation of that radical which was used in this work. Despite the possibilities for additive errors, it was found that for 90% of all the compounds studied, the difference between experimental and calculated BDE values, using the E_R values of the table, was only 3 kpm or smaller, the overall average difference being about 1.5 kpm. The remaining 10% showed differences greater than 3 but still not very large, and the possibility of experimental origin is not unreasonable. There are no consistencies apparent in the type of radical or compound, possibly excepting some hydroxyl and hydrogen compounds.

On the other hand, it seems likely that the bond dissociation energies of water are accurate, yet these suggest a reorganizational energy of 8 for hydroxyl whereas an average of about 10 is found for all the hydroxyl compounds studied. This discrepancy cannot easily be dismissed.

It seems reasonable to expect different reorganizational energies for the same radical when it is differently attached in the compound. For example, the nitro radical should have a different reorganizational energy in converting to NO_2 molecule, depending on whether it was bound through nitrogen or oxygen in the compound. The value of -14 in the table refers to nitrogen-bonded combinations. Data for oxygen-bonded combinations are too sparse and uncertain to be included

TABLE 10-1

The Application of Reorganizational Energies to Calculation of Bond Dissociation Energies

Radical	CBE (calc.)	$E_R(A,B)$	BDE (calc.)	BDE (exp.)
OC (−26.4)		−65		
O	192	−65, 0	127	127
S	139	−65, 0	74	74
SC (+56)		−42		
O	189	−42, 0	147	150
S	139	−42, 0	97	95
$C_6H_5'CO$ (19)		−15		
$C_6H_5'CH_2$	83	−15, −13	55	59
n-C_3H_7	85	−15, −1	69	71
C_2H_5	85	−15, −1	69	71
H	97	−15, 0	82	80
Cl	86	−15, 0	71	74
Br	73	−15, 0	58	58
OC_2H_5	110	−15, 1	91	88
CH_3	85	−15, 2	72	71
OH	110	−15, 10	105	98
C_6H_5'	85	−15, 15	85	84
CHO (−4)		−15		
CH_2CHCH_2	84	−15, −10	59	61
i-C_3H_7	83	−15, −2	66	66
n-C_3H_7	83	−15, −1	67	66
C_2H_5	84	−15, −1	68	68
H	92	−15, 0	77	74
CH_3O	108	−15, 1	94	94
CH_3	84	−15, 2	71	70
OH	112	−15, 10	107	96
C_6H_5'	85	−15, 15	85	85
-NO_2 (8.1)		−14		
NO_2	36	−14, −14	8	14
i-C_3H_7	74	−14, −2	58	60
n-C_3H_7	74	−14, −1	59	61
C_2H_5	74	−14, −1	59	58
F	60	−14, 0	46	45
Cl	55	−14, 0	41	34
CH_3	74	−14, 2	62	60
OH	53	−14, 10	49	49

TABLE 10-1 (continued)

Radical	CBE (calc.)	$E_R(A,B)$	BDE (calc.)	BDE (exp.)
$C_6H_5CH_2$ (45)		−13		
$C_6H_5CH_2$	83	−13, −13	57	58
SO_2CH_3	73	−13, −12	48	47
t-C_4H_9	83	−13, −5	65	64
CH_3CO	83	−13, −5	65	64
i-C_3H_7	83	−13, −2	68	68
n-C_3H_7	83	−13, −1	69	69
C_2H_5	83	−13, −1	69	69
H	99	−13, 0	86	85
Cl	83	−13, 0	70	70
Br	70	−13, 0	57	55
I	55	−13, 0	42	40
CH_3S	71	−13, 0	58	56
SH	71	−13, 1	59	56
OCH_3	82	−13, 1	70	70
CH_3	83	−13, 2	72	72
OH	81	−13, 10	78	78
NH_2	72	−13, 10	69	72
$CH_2{=}CH$	83	−13, 12	82	85
C_6H_5	85	−13, 15	87	88
CH_3CO (−5)		−13		
H	97	−13, 0	84	87
OCH_3	110	−13, 1	98	97
OC_2H_5	110	−13, 1	98	96
OH	110	−13, 10	107	107
NH_2	100	−13, 10	97	95
CH_3SO_2 (−63)		−12		
$C_6H_5CH_2$	73	−12, −13	49	51
CH_2CHCH_2	73	−12, −10	51	51
t-C_4H_9	73	−12, −5	56	56
i-C_3H_7	73	−12, −2	59	59
C_2H_5	73	−12, −1	60	61
CH_3	73	−12, 2	63	60
C_6H_5	73	−12, 15	76	78

TABLE 10-1 (continued)

Radical	CBE (calc.)	$E_R(A,B)$	BDE (calc.)	BDE (exp.)
CH_2CHCH_2 (41)		−10		
CHO	84	−10, −15	59	61
SO_2CH_3	73	−10, −12	51	51
t-C_4H_9	83	−10, −5	68	67
i-C_3H_7	83	−10, −2	71	71
n-C_3H_7	83	−10, −1	72	72
C_2H_5	83	−10, −1	72	72
H	99	−10, 0	89	88
Cl	83	−10, 0	73	70
Br	70	−10, 0	60	56
I	55	−10, 0	45	44
OCH_3	81	−10, 1	72	68
CH_3	83	−10, 2	75	75
OH	81	−10, 10	81	82
NH_2	72	−10, 10	72	75
$CH_2=CH$	83	−10, 12	85	84
C_6H_5	85	−10, 15	90	93
t-C_4H_9 (7)		−5		
$C_6H_5CH_2$	83	−5, −13	65	64
CH_3SO_2	73	−5, −12	56	56
CH_2CHCH_2	83	−5, −10	68	67
t-C_4H_9	83	−5, −5	73	68
CH_2OH	83	−5, −2	76	71
i-C_3H_7	83	−5, −2	76	74
n-C_3H_7	83	−5, −1	77	77
C_2H_5	83	−5, −1	77	78
H	97	−5, 0	92	89
Cl	83	−5, 0	78	80
Br	70	−5, 0	65	66
I	55	−5, 0	50	50
CH_3S	71	−5, 0	66	66
SH	71	−5, 1	67	67
CH_3	83	−5, 2	80	81
OH	81	−5, 10	86	91
NH_2	72	−5, 10	77	77
$CH=CH_2$	83	−5, 12	90	89
C_6H_5	85	−5, 15	95	93

TABLE 10-1 (continued)

Radical	CBE (calc.)	$E_R(A,B)$	BDE (calc.)	BDE (exp.)
CH_3CO (−5)		−5		
$C_6H_5CH_2$	83	−5, −13	65	64
i-C_3H_7	85	−5, −2	78	76
n-C_3H_7	85	−5, −1	79	78
C_2H_5	85	−5, −1	79	78
Cl	87	−5, 0	82	82
Br	73	−5, 0	68	68
I	54	−5, 0	49	51
CH_3	85	−5, 2	82	81
C_6H_5	85	−5, 15	95	95
CH_3CHOH (−18)		−4		
i-C_3H_7	83	−4, −2	77	75
n-C_3H_7	83	−4, −1	78	78
C_2H_5	83	−4, −1	78	78
H	99	−4, 0	95	90
CH_3	83	−4, 2	81	81
NO (21.6)		−4		
F	60	−4, 0	56	56
Br	33	−4, 0	29	29
N	119	−4, 0	115	115
CH_2OH (−6)		−2		
i-C_3H_7	83	−2, −2	79	80
n-C_3H_7	83	−2, −1	80	81
C_2H_5	83	−2, −1	80	81
H	99	−2, 0	97	94
CH_3	83	−2, 2	83	84
C_6H_5	84	−2, 15	97	96

TABLE 10-1 (continued)

Radical	CBE (calc.)	$E_R(A,B)$	BDE (calc.)	BDE (exp.)
i-C_3H_7 (18)		-2		
CHO	83	$-2, -15$	66	66
NO_2	74	$-2, -14$	58	60
$C_6H_5CH_2$	83	$-2, -13$	68	68
CH_3SO_2	73	$-2, -12$	59	59
CH_2CHCH_2	83	$-2, -10$	71	71
CH_3CO	85	$-2, -5$	78	76
i-C_3H_7	83	$-2, -2$	79	79
n-C_3H_7	83	$-2, -1$	80	81
C_2H_5	83	$-2, -1$	80	81
H	98	$-2, 0$	96	95
Cl	83	$-2, 0$	81	81
Br	70	$-2, 0$	68	69
I	55	$-2, 0$	53	54
CH_3S	71	$-2, 0$	69	70
HS	71	$-2, 1$	70	70
OCH_3	81	$-2, 1$	80	81
CH_3	83	$-2, 2$	83	84
CH_3COO	80	$-2, 6$	84	88
OH	81	$-2, 10$	89	92
NH_2	72	$-2, 10$	80	79
$CH_2{=}CH$	83	$-2, 12$	93	93
C_6H_5	85	$-2, 15$	98	97
n-C_3H_7 (21)		-1		
C_6H_5CO	85	$-1, -15$	69	71
CHO	83	$-1, -15$	67	66
NO_2	74	$-1, -14$	59	61
$C_6H_5CH_2$	83	$-1, -13$	69	69
CH_2CHCH_2	83	$-1, -10$	72	72
t-C_4H_9	83	$-1, -5$	77	77
CH_3CO	85	$-1, -5$	79	78
CH_3CHOH	83	$-1, -4$	78	78
CH_2OH	83	$-1, -2$	80	81
i-C_3H_7	83	$-1, -2$	80	81
n-C_3H_7	83	$-1, -1$	81	82
C_2H_5	83	$-1, -1$	81	82
H	99	$-1, 0$	98	98
Cl	83	$-1, 0$	82	81
Br	70	$-1, 0$	69	69
I	55	$-1, 0$	54	54
CH_3S	71	$-1, 0$	70	71
SH	71	$-1, 1$	71	71
CH_3	83	$-1, 2$	84	85
OH	81	$-1, 10$	90	92
NH_2	72	$-1, 10$	81	79
$CH_2{=}CH$	83	$-1, 12$	94	94
C_6H_5	85	$-1, 15$	99	99

TABLE 10-1 (continued)

Radical	CBE (calc.)	$E_R(A,B)$	BDE (calc.)	BDE (exp.)
C_2H_5 (26)		-1		
C_6H_5CO	85	$-1, -15$	69	71
CHO	84	$-1, -15$	68	68
NO_2	74	$-1, -14$	59	58
$C_6H_5CH_2$	83	$-1, -13$	69	69
CH_3SO_2	73	$-1, -12$	60	61
CH_2CHCH_2	83	$-1, -10$	72	72
CH_3CO	84	$-1, -5$	78	78
t-C_4H_9	83	$-1, -5$	77	78
CH_3CHOH	83	$-1, -4$	78	78
i-C_3H_7	83	$-1, -2$	80	81
CH_2OH	83	$-1, -2$	80	81
n-C_3H_7	83	$-1, -1$	81	82
C_2H_5	83	$-1, -1$	81	82
H	99	$-1, 0$	98	98
Cl	83	$-1, 0$	82	81
Br	70	$-1, 0$	69	68
I	55	$-1, 0$	54	54
CH_3S	71	$-1, 0$	70	70
OCH_3	82	$-1, 1$	82	81
OC_2H_5	82	$-1, 1$	82	81
CH_3	83	$-1, 2$	84	85
SH	71	$-1, 1$	71	71
OH	81	$-1, 10$	90	91
NH_2	72	$-1, 10$	81	78
$CH_2=CH$	83	$-1, 12$	94	94
C_6H_5	85	$-1, 15$	99	99
CH_3S (30)		0		
$C_6H_5CH_2$	71	$0, -13$	58	56
t-C_4H_9	71	$0, -5$	66	66
i-C_3H_7	71	$0, -2$	69	70
n-C_3H_7	71	$0, -1$	70	71
C_2H_5	71	$0, -1$	70	70
H	88	$0, 0$	88	87
CH_3	71	$0, 2$	73	73
C_6H_5	73	$0, 15$	88	86

TABLE 10-1 (continued)

Radical	CBE (calc.)	$E_R(A,B)$	BDE (calc.)	BDE (exp.)
HS (34)		1		
$C_6H_5CH_2$	71	1, −13	59	56
t-C_4H_9	71	1, −5	67	67
i-C_3H_7	71	1, −2	70	70
n-C_3H_7	71	1, −1	71	71
C_2H_5	71	1, −1	71	71
H	88	1, 0	89	91
CH_3	71	1, 2	74	73
C_6H_5	73	1, 15	89	87
CH_3O (3)		1		
CHO	108	1, −15	94	94
CH_3CO	110	1, −13	98	97
CH_2CHCH_2	81	1, −10	72	68
i-C_3H_7	81	1, −2	80	81
n-C_3H_7	81	1, −1	81	81
C_2H_5	81	1, −1	81	81
H	113	1, 0	114	103
OCH_3	33	1, 1	35	36
CH_3	82	1, 2	85	81
OH	33	1, 10	44	43
C_6H_5	82	1, 15	98	100
C_2H_5O (−5)		1		
C_6H_5CO	110	1, −15	96	88
CH_3CO	110	1, −13	98	96
t-C_4H_9	81	1, −5	77	80
i-C_3H_7	81	1, −2	80	81
C_2H_5	82	1, −1	82	81
H	113	1, 0	114	103
OC_2H_5	33	1, 1	35	34
CH_3	82	1, 2	85	81
OH	33	1, 10	44	43
C_6H_5	85	1, 15	101	101

TABLE 10-1 (continued)

Radical	CBE (calc.)	$E_R(A,B)$	BDE (calc.)	BDE (exp.)
CH_3 (34)		2		
C_6H_5CO	85	2, −15	72	74
CHO	84	2, −15	71	70
NO_2	74	2, −14	62	60
$C_6H_5CH_2$	83	2, −13	72	72
CH_3SO_2	73	2, −12	63	60
CH_2CHCH_2	83	2, −10	75	75
CH_3CO	84	2, −5	81	81
$t\text{-}C_4H_9$	83	2, −5	80	81
CH_3CHOH	83	2, −4	81	81
$i\text{-}C_3H_7$	83	2, −2	83	84
CH_2OH	83	2, −2	83	84
$n\text{-}C_3H_7$	83	2, −1	84	85
C_2H_5	83	2, −1	84	85
H	99	2, 0	101	104
Cl	83	2, 0	85	84
Br	70	2, 0	72	70
I	55	2, 0	57	56
CH_3S	71	2, 0	73	73
CH_3O	82	2, 1	85	81
C_2H_5O	82	2, 1	85	81
CH_3	83	2, 2	87	88
SH	71	2, 1	74	73
CH_3COO	80	2, 6	88	87
OH	81	2, 10	93	91
NH_2	71	2, 10	83	80
$CH_2\text{=}CH$	83	2, 12	97	97
C_6H_5	85	2, 15	102	102
CH_3COO (−45)		6		
$i\text{-}C_3H_7$	81	6, −2	85	88
C_2H_5	81	6, −1	86	87
H	113	6, 0	119	111
CH_3	81	6, 2	89	87
$CH_2\text{=}CH$	81	6, 12	99	99

TABLE 10-1 (continued)

Radical	CBE (calc.)	$E_R(A,B)$	BDE (calc.)	BDE (exp.)
OH (9)		10		
C_6H_5CO	110	10, −15	105	98
CHO	112	10, −15	107	96
NO_2	53	10, −14	49	49
$C_6H_5CH_2$	81	10, −13	78	78
CH_3CO	110	10, −13	107	107
CH_2CHCH_2	81	10, −10	81	82
$t\text{-}C_4H_9$	81	10, −5	86	91
$i\text{-}C_3H_7$	81	10, −2	89	92
$n\text{-}C_3H_7$	81	10, −1	90	92
C_2H_5	81	10, −1	90	91
H	111	10, 0	121	119
Cl	49	10, 0	59	60
Br	49	10, 0	57	56
I	54	10, 0	64	56
CH_3O	33	10, 1	44	43
C_2H_5O	33	10, 1	44	43
CH_3	81	10, 2	93	91
OH	33	10, 10	53	51
C_6H_5	85	10, 15	110	112
NH$_2$ (41)		10		
$C_6H_5CH_2$	72	10, −13	69	72
CH_3CO	100	10, −13	97	96
$t\text{-}C_4H_9$	72	10, −5	77	77
$i\text{-}C_3H_7$	72	10, −2	80	79
$n\text{-}C_3H_7$	72	10, −1	81	79
C_2H_5	72	10, −1	81	78
H	93	10, 0	103	104
CH_3	72	10, 2	84	81
NH_2	39	10, 10	59	59
C_6H_5	72	10, 15	97	100
$CH_2{=}CH$ (68)		12		
$C_6H_5CH_2$	83	12, −13	82	85
CH_2CHCH_2	83	12, −10	85	84
$t\text{-}C_4H_9$	83	12, −5	90	89
$i\text{-}C_3H_7$	83	12, −2	93	92
$n\text{-}C_3H_7$	83	12, −1	94	94
C_2H_5	83	12, −1	94	94
H	99	12, 0	111	108
Cl	86	12, 0	98	88
CH_3O	81	12, 1	94	93
C_2H_5O	81	12, 1	94	97
CH_3	83	12, 2	97	97
$CH_2{=}CH$	83	12, 12	107	110
C_6H_5	85	12, 15	112	113

TABLE 10-1 (continued)

Radical	CBE (calc.)	$E_R(A,B)$	BDE (calc.)	BDE (exp.)
C_6H_5 (80)		15		
C_6H_5CO	85	15, −15	85	84
CHO	85	15, −15	85	85
$C_6H_5CH_2$	85	15, −13	87	88
CH_3SO_2	73	15, −12	76	78
CH_2CHCH_2	85	15, −10	90	93
CH_3CO	85	15, −5	95	95
t-C_4H_9	85	15, −5	95	93
CH_2OH	85	15, −2	98	98
i-C_3H_7	85	15, −2	98	97
n-C_3H_7	85	15, −1	99	99
C_2H_5	85	15, −1	99	99
H	97	15, 0	112	112
Cl	87	15, 0	102	97
Br	71	15, 0	86	82
I	57	15, 0	72	66
CH_3S	73	15, 0	88	86
SH	73	15, 1	89	87
CH_3O	85	15, 1	101	100
C_2H_5O	85	15, 1	101	101
CH_3	85	15, 2	102	102
CH_3COO	85	15, 6	106	102
OH	85	15, 10	110	112
NH_2	72	15, 10	97	100
CH_2=CH	85	15, 12	112	113
C_6H_5	85	15, 15	115	116

in the table, but it appears that E_R for these is approximately double the value for the nitrogen-bonded compounds. In other words, the −ONO radical in combination appears less stable than the −NO_2 radical in that the conversion of the former to NO_2 molecule is more highly exothermic.

There is one example in the table of a radical exhibiting two different reorganizational energies, −5 for a group of 9 compounds and −13 for another group of 5 compounds. This is the acetyl radical, CH_3CO. The phenomenon appears to be real, although perhaps it is not. One may speculate about the possibility of isomeric free radicals, perhaps CH_2=C(OH) and CH_3C=O. But other differences for a given radical appear quite random and are probably due to error.

We are ill equipped to speculate about the steric factor except in a very qualitative manner. But we should keep in mind that contributions to the reorganizational energy may arise from spacial requirements of the components of the radical. This point becomes very important in the discussion of molecular addition compounds later in this chapter.

SPECULATIVE EXPLANATIONS OF
REORGANIZATIONAL ENERGY

Perhaps one day a consistent theory of reorganizational energy may become available so that a trained chemist can look at the composition and structure of a given radical and predict with assurance its reorganizational energy. Meanwhile we can only delve about in a fragmentary manner, choosing isolated examples which appear to expose some hint of their nature.

The contrast between phenyl and benzyl provides a possible illustration of the stabilizing influence of the π electron system in the aromatic ring. Removal of one hydrogen atom from a benzene molecule appears to upset the "resonance" to the extent of weakening the residual bonding by 15 kpm. We need to understand how the mere release of one carbon electron from its previous involvement with a bond to hydrogen (or to any other substituent) could create such havoc in the ring. On the other hand, release of a hydrogen atom from the methyl group of toluene seems to allow the methylene residue to enter into the aromatic system to the extent of strengthening the residual bonding by 13 kpm. Again, no easy description of how this comes about appears to be available unless in terms of "resonance forms."

Particular interest attaches to the effect of substituting methyl for hydrogen on methane. The reorganizational energy of methyl is +2 kpm, that of ethyl, −1, isopropyl, −2, and tertiary butyl, −5 kpm. This replacement of a C–H bond by a C–C bond may be imagined as having several effects that might influence the residual bonding. First, it may reduce repulsions somewhat by allowing more space for the bonding electrons, carbon atoms being larger than hydrogen atoms. Second, it substitutes a nonpolar C–C bond for a slightly polar C–H bond that crowds the bonding electrons around the carbon enough to increase repulsions somewhat. Such effects appear to involve only direct atomic contact, so that a larger alkyl group has no more effect than methyl. But such factors could cause the observed trend in E_R values, which are of course highly significant in their influence on relative reactivities of primary, secondary, and tertiary positions, wherein scarcely any light is shed by CBE values. A similar "trend" is noted in the reorganizational energies of the alcohol radicals: CH_2OH, −2 kpm; and CH_3CHOH, −4 kpm.

We have already noted the dramatic effect of a lone pair of outermost electrons in weakening the homonuclear single covalent bond energy. It seems possible that liberation of a single electron from a bond might have a similar effect on the residual bonding in the radical, weakening it still more. For example, this might account for the reorganizational energy of about 8, or 10, for the hydroxyl radical.

Although the carboxyl radical, −COOH, is not included in Table 10-1 because of paucity of data, the available evidence suggests that it has an E_R value close to zero. Evidently the liberated bonding orbital here can have little effect. However, the splitting of a hydrogen atom from this −COOH radical can be predicted to be favored by the ability of the residual fragment to rearrange to a stable molecule of carbon dioxide. The total C–O energy in –COOH, calculated as CBE, is about 173 + 110 = 283 kpm, and the O–H energy about 113 kpm. But the gain from 283 to 384 in CO_2 is 101 kpm, allowing the BDE for the O–H bond to be only 12 kpm. The experimental value is indeed 12 kpm.

Other BDE values for radicals can similarly be accounted for. For example, removal of a methyl hydrogen from methanol leaves –CH_2OH, involving apparently an E_R of about –2 kpm. That is, CBE = 99 but BDE = 97. If now we split off the hydroxyl hydrogen, the residue can rearrange to formaldehyde, HCHO. The CBE for the O–H bond is 111 kpm, but the rearrangement releases 85 kpm. Thus the BDE is 111 − 85 + 2 = 28 kpm. The experimental value is reported to be 31 kpm.

There is also the interesting example of acetic acid, CH_3COOH, from which the splitting of the hydroxyl hydrogen appears to cost about 6 kpm more than the CBE, or a total BDE of 119. The bonding in the acetate radical has been weakened by about 6 kpm, but we have no way of knowing exactly where or how. However, it is clear that splitting off a methyl group would allow formation of a molecule of carbon dioxide, with 95–101 kpm evolved. The CBE for the C–C bond, before the splitting off of the acetate radical, was 81, so it cannot be much smaller in the radical. The E_R for CH_3– is 2, so that the BDE is about 78 + 2 − 98 = −18 kpm. The "experimental" value is −20 kpm. This shows the acetate radical to be exothermically decomposed to methyl radical and carbon dioxide.

As a final example we may consider the extraordinarily high chemical reactivity of ozone, in which the average bond energy is 72 kpm. Breaking loose one oxygen atom, however, will permit the residue to form a very stable O_2 molecule, in which the bond energy is 119 kpm. This gain of 47 kpm in stability is reflected in the great ease with which an atom of oxygen is liberated from O_3, for the BDE is only 72 − 47 = 25 kpm.

GENERAL APPLICATIONS OF BDE

Small differences in bond strengths have been found to make very large contributions to relative rates of competitive reactions. Hence even small differences, in E_R and therefore BDE, may be highly significant in reactions involving free radicals. Special attention is therefore directed to the results discussed briefly above that distinguish primary from secondary from tertiary reactivities. The CBE is indeed the principal factor determining whether the free energy of a chemical change is negative, but a knowledge of BDE may be very important in understanding rates, mechanisms, and distribution of products, when several may represent nearly the same total energy.

This type of approach may also be useful in the study of ionic reactions, as in determining and understanding the relative stabilities of carbonium ions.

The effect of ionization of a molecule or radical on the residual bond strength is also closely related. For example, the energy of the N–N bond in N_2 is 226 kpm,

TABLE 10-2
Tentative Reorganizational Energies of
Radicals, Supplementing Table 10-1

Radical	E_R	Radical	E_R
OCl	−35	c-C_5H_9	−5
C_6H_5O	−29	c-C_6H_{11}	−5
HOO	−22	C_2H_5COO	−4
$C_6H_5NCH_3$	−20	c-C_4H_7	−4
C_6H_5S	−14	i-C_3H_7O	0
$(C_6H_5)_3C$	−14	OBO	1
n C_3H_7COO	−11	c-C_3H_5	2
AlF_2	−9	OF	7
$AlCl_2$	−6	BO	19
NF_2	−6	GeH_3	19

but for the N_2^+ molecule, dissociation to $N + N^+$ costs only 200 kpm. The difference is easily shown to be that between ionizing N_2, 361 kpm, and ionizing N, 335 kpm, or 26 kpm.

Clearly, much work lies ahead, but at least we have progressed to the point of recognizing, and in many instances quantitatively understanding, how bond dissociation energy is based on the contributing bond energy even where they are very different. Concerning the assumed constancy of radical reorganizational energy, it can only be said at this time that in most examples, if other factors not yet recognized or evaluated also influence the reorganizational energies so that they are not constant from compound to compound for a given radical, such factors have effects that are in magnitude within the limits of experimental measurement of bond dissociation energies. The constancy assumed for reorganizational energies appears strongly supported by the data available to date. The listing in Table 10-1 is extended in Table 10-2, which suggests E_R values for additional radicals based on only a small amount of evidence and therefore to be used only in a tentative or exploratory manner.

COORDINATION IN MOLECULAR ADDITION COMPOUNDS

What is commonly called a coordinate covalent bond is any bond in which the two bonding electrons are provided by one atom and a vacant orbital by the other. The usual understanding appears to be that, once formed, such a bond is a two electron bond just like another two electron covalent bond.

A most difficult problem is that of acquiring a quantitative understanding of coordinate covalence. Consider two atoms sharing an electron pair with each other, while each is also covalently bonded to other atoms. If this bond was formed by an electron contributed by each atom of the bonded pair, then homolytic bond rupture is to be expected, and the two fragments will be neutral free radicals. But if one of the atoms originally contributed both electrons of the bond and the other atom contributed none, then the expected bond rupture will be heterolytic, forming two neutral molecules, a donor and an acceptor. An immediate question is, does this difference affect the contribution made by the coordinate covalent bond to the total atomization energy of the molecular addition compound?

A closely related question concerns the electronegativity equalization concept. It is easy to imagine the equalization of electronegativities through transfer of electrons from donor to acceptor. But what if the acceptor is less electronegative than the donor as, for example, in $(CH_3)_3BN(CH_3)_3$? The partial charges are: C, -0.059; H, O, and B, 0.174 in $B(CH_3)_3$; and C, -0.030, H, 0.031, N, -0.186 in $N(CH_3)_3$. But in the addition compound, assuming electronegativity equalization, the charges are: C, -0.044; H, 0.015; B, 0.194; and N, -0.197. In effect, a transfer of 0.195 electron *from acceptor to donor* must occur.

Another problem arises from the steric factors of coordination, especially the result of structural changes in the acceptor that are forced by coordination. For example, a planar M3 acceptor molecule must become tetrahedral in coordination. This brings the substituents closer not only to each other, but also to the various parts of the donor molecule, creating increased stresses that must weaken the overall bonding.

To these fundamental problems may be added the paucity of experimental data with which to test hypotheses. The following discussion is of necessity speculative, but it is appropriate to this chapter because the donor–acceptor interactions in molecular addition compounds are experimentally measurable only as bond dissociation energies.

The question of whether the coordination bond can be treated as an ordinary contributing bond energy cannot yet be decisively answered. On the one hand, one might suppose the CBE to be independent of the source of the bonding electrons.

For example, the atomization energy of $BF_3 \cdot N(CH_3)_3$ is calculated to be 1593 kpm, compared to the experimental 1592 kpm. The CBE of the B–N coordination bond is 82.9 kpm, but the experimental BDE is only 28.6 kpm, implying that the E_R totals 54.3 kpm. In fact, the coordination causes the B–F bonds, becoming tetrahedral from planar, to lengthen and weaken by a total of 463 $-410 = 53$, thus accounting for essentially the entire reorganizational energy.

Unfortunately, data for other molecular addition compounds are sparse and, where available, do not lead to as satisfactory results. The calculated and experimental total atomization energies of $BF_3 \cdot O(CH_3)_2$ and $BF_3 \cdot O(C_2H_5)_2$, for example, are 1263–1235, and 1815–1801 kpm. For $BF_3 \cdot P(CH_3)_3$, they are 1580 and 1562 kpm. The calculated CBE for the coordination bond B–O is 96.9 but the BDE only about 12, indicating an E_R of 85 kpm. Only 49 kpm is accounted for by B–F bond weakening. BDEs of 29 kpm for $BF_3N(CH_3)_3$ but only 12 for $BF_3O(CH_3)_2$ seem incongruous.

On the other hand, Drago and co-workers[4] have found an extraordinary relationship in which two constant parameters, C and E, can be assigned to each donor or acceptor molecule such that the enthalpy of formation of gaseous AB from gaseous monomeric A and B is given by

$$\Delta H = C_A C_B + E_A E_B$$

It seems rather surprising that the same parameters are applicable to combinations whether they involve protonic bridging, charge transfer complexes, or electron pair coordination. It hardly seems possible that such a relationship could have its reportedly high validity if the coordinate covalent bond is really like an ordinary covalent bond. This success suggests that the coordination bond may be principally electrostatic.

In a recent study I have tentatively *assumed* that the coordinate covalent bonds *do* contribute to the total atomization energy as if they were normal polar covalent bonds. A selection of seven Lewis acids and thirty bases allowed calculation, using the Drago parameters, of 210 combinations. The average calculated CBE for these 210 one-to-one molecular addition compounds is about 69 kpm. In contrast, the average calculated BDE is only 18 kpm. This would indicate E_R values, acid and base combined, approximating around 50 kpm. The steric strains, called B− and F− by H. C. Brown,[5] would surely be involved in these E_R values, but such high E_R values seem questionable when no fundamental change in the bonding is likely to accompany the release of the Lewis acid and base from their coordination.

Among the bases included in this study, three types are well represented: eight amines, nine oxygen compounds, and five organic sulfides. The following principal observations were made:

(1) With a given acid, E_R is greater for O, next for S, and least for N ligands.

(2) The E_R order for the seven acids is the same for all three base types, decreasing as follows: $B(CH_3)_3$, BF_3, $Al(C_2H_5)_3$, $Al(CH_3)_3$, $Ga(C_2H_5)_3$, $Ga(CH_3)_3$, and $In(CH_3)_3$.

We can account qualitatively for the order in (2) above in terms of the crowding around the central atom decreasing as the central atom becomes larger. In the exceptional example of aluminum and gallium which are very similar in radius, aluminum is much less electronegative and therefore more positive in analogous compounds. Being more positive, it is smaller for this reason. The substituents thus come closer together, and they are also more negative and hence exert greater repulsions.

Table 10-3 summarizes these results. Remember that they are all based on what may be a faulty assumption regarding the nature of the coordination bond.

[4] R. S. Drago, G. C. Vogel, and T. E. Needham, *J. Am. Chem. Soc.* **93**, 6014 (1971).
[5] H. C. Brown, *J. Am. Chem. Soc.* **67**, 374, 378 (1945).

TABLE 10-3

Assumed Reorganization Energies (kcal mol^{-1}) in Decreasing Order

Ether and carbonyl oxygen		Sulfide sulfur		Amine nitrogen	
$B(CH_3)_3$	83				
BF_3	80				
$Al(C_2H_5)_3$	69				
$Al(CH_3)_3$	67				
				$B(CH_3)_3$	65
		$B(CH_3)_3$	64		
		BF_3	63		
				BF_3	60
		$Al(CH_3)_3$	57		
$Ga(C_2H_5)_3$	55	$Al(C_2H_5)_3$	55		
$Ga(CH_3)_3$	53				
$In(CH_3)_3$	52				
		$Ga(C_2H_5)_3$	51		
		$Ga(CH_3)_3$	49	$Al(C_2H_5)_3$	49
		$In(CH_3)_3$	49	$Al(CH_3)_3$	49
				$Ga(C_2H_5)_3$	42
				$Ga(CH_3)_3$	39
				$In(CH_3)_3$	38

In the perhaps more likely event that the coordination bond is indeed very different from the ordinary covalent bond, there appears to be no easy route to further insight because of the difficulty of measuring E_R or understanding its quantitative variations. An adequate method of CBE calculation might be improvised but with no firm basis for testing its accuracy. The BDE is a measureable quantity, but it is the sum of two undeterminable variables.

ELEVEN

Chemical Bonds
in Organic Compounds

THE NATURE OF THE PROBLEM

Even though most of my industrial research efforts were spent on the investigation of organic compounds and their reactions, it has never been my privilege to become a real, honest-to-goodness organic chemist. At the outset of this chapter, therefore, I wish to avoid any hint of false pretense by admitting my limitations in this area even before they become obvious. On the other hand, some years ago I felt impelled to request that my academic title be changed from "Professor of Inorganic Chemistry" to "Professor of Chemistry." This was simply because I felt the former title to imply boundaries which might cramp my style. Theoretical chemistry, with its limitless dependence on complex mathematics, also seemed a distant and alien field. This left me in the highly vulnerable position of generalist, wherein the primary requisite, omniscience, lies well beyond mortal reach. Nevertheless I have found the role fascinating, because the overall view of chemistry is in many ways, I believe, the most appealing.

Organic chemistry was in fact my first love. When, as a high school student, I first became acquainted with chemistry, it was Edwin Slosson's "Creative Chemistry" that really seduced me into this profession, saving the world from one more starving artist. When, two months later, it became perfectly clear that already I knew far more chemistry than my high school teacher (after all, I had studied the subject two months longer than he), the future seemed more assured than ever thereafter. And when, as a college senior, I defined physical chemistry as the triumph of matter over mind, it was my erstwhile instructor who assured me, with a fiendish twinkle in his eye, "Well, it was never much of a triumph in your case, was

177

it!" Indeed it was only a quirk of Fate, in the shape of The Great Depression, that found me working at an inorganic chemical plant, where my interest in inorganic chemistry was first created.

If there is a legitimate point to all of this seemingly irrelevant reminiscing, it lies in the fact that my diversity of experience has led to a viewpoint that probably differs from that of the average specialist. Consequently I find it disturbing that specialists in different areas have developed different jargon and different ways of thinking that inhibit communication among them. I suspect there is much they could otherwise learn from each other. In particular it seems unfortunate that organic chemists seem to live so much in their own world, when carbon is really not conspicuously more individualistic than other elements.

It has long seemed to me that the "uniqueness" of organic chemistry is derived not from the uniqueness of carbon atoms but from the uniqueness of the combination of carbon atoms with hydrogen atoms. On the basis of this conviction, I have shunned thinking of limestone as an organic compound and have much preferred to define organic compounds as those having at least one C—H bond per molecule. In this light the oft-repeated tale of the Wohler synthesis takes on an amusing aspect. It is alleged to have proved, and indeed it did prove convincing, that "vital force" is not necessary to produce an organic compound. According to the above definition, which in a logical sense seems superior to others, neither ammonium cyanate nor urea is an organic compound, so the Wohler synthesis became famous through the lucky accident of misinterpretation.

Carbon and hydrogen happen to be nearly equal in electronegativity, the carbon being slightly more attractive to electrons. This causes C—H bonds to be only slightly polar, which in turn means that the polarity of one bond will not appreciably affect the other bonds formed by the same carbon atom. Thus carbon atom chains and networks are easily formed in combination with hydrogen. Carbon atoms are also near enough to hydrogen atoms in size for strong bonding. Perhaps most important, carbon and hydrogen are the only elements of the entire periodic system whose atoms, when they have formed all of the single covalent bonds of which they are capable, have neither outer vacancies nor outer electron pairs which might be employed in supplementary bonding. This means that saturated C—H groups are capable of resisting chemical attack under ordinary conditions and thus moving "unchanged" through a great variety of reactions.

It is also worth noting that, although carbon and hydrogen are both a little above the median in electronegativity, they are less electronegative than most of the other nonmetals. Therefore they become partially positive in combination with the other nonmetals or groups of nonmetal atoms. This corresponds to partial negative charge on the other nonmetal, which turns out to be the chemically reactive part of the hydrocarbon derivative or functional group. It is instructive to think of the **functional group as the negatively charged portion of the molecule**. In fact, when the hydrocarbon group becomes partially negative, as for example in metal alkyls, then it serves as the functional group.

The marvelous scope and diversity of organic chemistry arise from the special nature of combinations of carbon with hydrogen together with the ability of the carbon to form additional bonds to nitrogen, oxygen, halogen, sulfur, and other elements.

The very nature of hydrocarbons and their derivatives places a great strain on any system of analyzing and calculating bond energies. Possibly the two principal causes are (1) steric and (2) number of bonds per molecule. Even in pure hydrocarbons, interactions between nonbonded atoms may influence significantly the molecular stability because of steric repulsions. Furthermore, for a molecule like HCl we can tolerate an error of about half a kilocalorie in the bond energy, but where any error must be multiplied by the number of bonds, the result may be less than satisfactory. For example, an error of 0.5 kpm in the C—H energy would produce a 5 kpm error in the atomization energy of butane. Therefore minor factors or slight errors cannot be tolerated or ignored as easily for most organic compounds as they commonly can in inorganic compounds.

There is abundant experimental evidence that C—H single covalent bonds are not all alike, nor are C—C single covalent bonds. Consider, for instance, the data for 75 isomeric decanes for which the standard heats of formation vary between extremes by more than 10 kpm. In each decane molecule of whatever isomer, there are the same number of C—C bonds and the same number of C—H bonds. By the methods described earlier in this book, all isomers would be assumed to have exactly the same bond strength and therefore the identical atomization energy. There have been numerous, often amazingly successful attempts to assign additive parameters to individual bonds and groups such that their sum correctly represents the experimental standard heat of formation (or atomization energy).[1] However, these have been completely empirical and shed no light on the fundamental causes of differences.

I have found that the atomization energies of a considerable number of alkanes can be represented quite accurately by addition of the appropriate bond energies chosen from Table 11-1. The data of Table 11-2 illustrate the possibilities but omit those isomers in which steric interferences of nearest or next to nearest methyl groups appear to weaken the bonding significantly. The same empirical bond energies listed in Table 11-1, unfortunately, do not serve well for derivatives of these hydrocarbons.

[1] C. T. Zahn, *J. Chem. Phys.* 2, 671 (1934); J. R. Platt, *J. Chem. Phys.* 15, 419 (1947); J. R. Platt, *J. Phys. Chem.* 56, 328 (1952); R. D. Brown, *J. Chem. Soc.* 2615 (1953); M. J. S. Dewar and R. Pettit, *J. Chem. Soc.* 1625 (1954); K. J. Laidler, *Can. J. Chem.* 34, 626 (1956); J. B. Greenshields and F. D. Rossini, *J. Phys. Chem.* 62, 271 (1958); T. L. Allen, *J. Chem. Phys.* 31, 1039 (1959); E. G. Lovering and K. J. Laidler, *Can. J. Chem.* 38, 2367 (1960); V. M. Tatevskii, V. A. Benderskii, and S. S. Yarovoi, "Rules and Methods for Calculating the Physico-Chemical Properties of Paraffinic Hydrocarbons," (translation, B. P. Mullins, ed.) Pergamon Press, Oxford, 1961; H. A. Skinner and G. Pilcher, *Quart. Rev.* 17, 264 (1963). Also J. W. Anderson, G. H. Beyer, and K. M. Watson, *Nat. Petrol. News* 36, R 476 (1944); J. L. Franklin, *Ind. Eng. Chem.* 41, 1070 (1949); M. Souders, C. S. Mathews, and C. O. Hurd, *Ind. Eng. Chem.* 41, 1408 (1949); S. W. Benson and J. H. Buss, *J. Chem. Phys.* 29, 546 (1958).

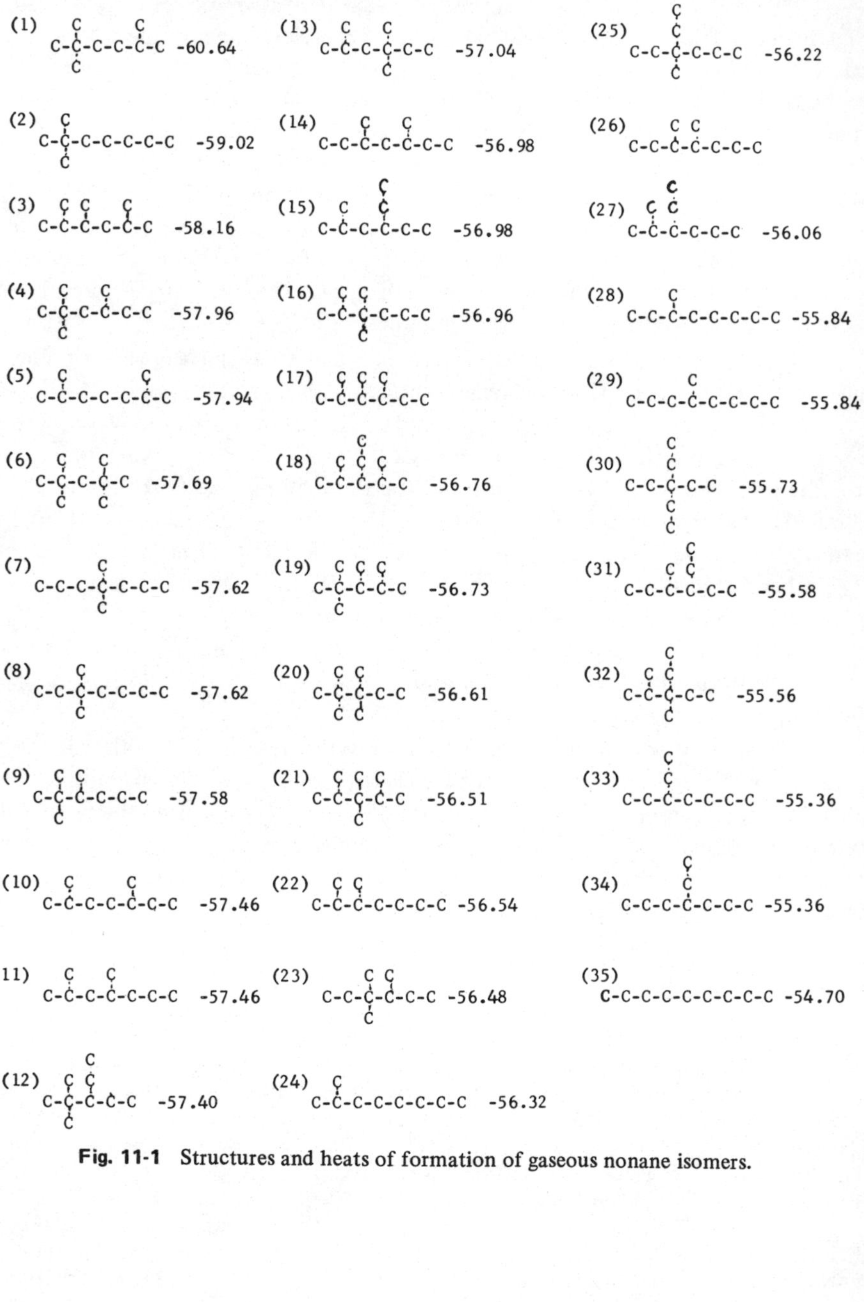

(1)
```
    C        C
C-C-C-C-C-C  -60.64
    C
```

(2)
```
    C
C-C-C-C-C-C-C  -59.02
    C
```

(3)
```
  C C   C
C-C-C-C-C-C  -58.16
```

(4)
```
  C   C
C-C-C-C-C-C  -57.96
    C
```

(5)
```
  C         C
C-C-C-C-C-C-C  -57.94
```

(6)
```
  C   C
C-C-C-C-C  -57.69
    C   C
```

(7)
```
         C
C-C-C-C-C-C-C  -57.62
         C
```

(8)
```
       C
C-C-C-C-C-C-C  -57.62
       C
```

(9)
```
  C C
C-C-C-C-C-C  -57.58
    C
```

(10)
```
  C     C
C-C-C-C-C-C-C  -57.46
```

11)
```
  C   C
C-C-C-C-C-C-C  -57.46
```

(12)
```
      C
  C C
C-C-C-C-C  -57.40
    C
```

(13)
```
  C   C
C-C-C-C-C-C  -57.04
      C
```

(14)
```
  C   C
C-C-C-C-C-C-C  -56.98
```

(15)
```
      C
  C   C
C-C-C-C-C-C  -56.98
```

(16)
```
  C C
C-C-C-C-C-C  -56.96
      C
```

(17)
```
  C C C
C-C-C-C-C-C  
```

(18)
```
  C C C
C-C-C-C-C  -56.76
```

(19)
```
  C C C
C-C-C-C-C  -56.73
    C
```

(20)
```
  C C
C-C-C-C-C  -56.61
    C C
```

(21)
```
  C C C
C-C-C-C-C  -56.51
      C
```

(22)
```
  C C
C-C-C-C-C-C-C  -56.54
```

(23)
```
    C C
C-C-C-C-C-C  -56.48
      C
```

(24)
```
  C
C-C-C-C-C-C-C  -56.32
```

(25)
```
      C
      C
C-C-C-C-C-C  -56.22
      C
```

(26)
```
    C C
C-C-C-C-C-C-C
```

(27)
```
      C
  C C
C-C-C-C-C-C  -56.06
```

(28)
```
      C
C-C-C-C-C-C-C-C  -55.84
```

(29)
```
        C
C-C-C-C-C-C-C-C  -55.84
```

(30)
```
      C
      C
C-C-C-C-C  -55.73
      C
      C
```

(31)
```
      C
  C C
C-C-C-C-C-C  -55.58
```

(32)
```
      C
  C C
C-C-C-C-C  -55.56
      C
```

(33)
```
      C
      C
C-C-C-C-C-C-C  -55.36
```

(34)
```
      C
      C
C-C-C-C-C-C-C  -55.36
```

(35)
```
C-C-C-C-C-C-C-C-C  -54.70
```

Fig. 11-1 Structures and heats of formation of gaseous nonane isomers.

TABLE 11-1
Empirical Bond Energies
Applicable to Alkanes

Bond type	Energy (kpm)
C–H'	98.5
C–H''	98.2
C–H'''	97.1
C'–C'	84.4
C'–C''	84.1
C'–C'''	84.7
C'–C''''	84.8
C''–C''	84.1
C''–C'''	84.4
C''–C''''	84.8
C'''–C'''	84.6

TABLE 11-2
Atomization Energies of Some Alkanes

Compound	Atomization energy (kpm)	
	Calc.	Exp.
Ethane	675.4	675.4
Propane	955.5	955.5
n-Butane	1236.1	1236.1
Isobutane	1237.6	1237.6
n-Pentane	1516.6	1516.6
Isopentane	1517.9	1518.5
Neopentane	1521.3	1521.3
n-Hexane	1797.1	1797.0
2-Methylpentane	1798.4	1798.7
2,2-Dimethylbutane	1801.5	1801.4
3-Methylpentane	1798.1	1798.1
2,3-Dimethylbutane	1799.6	1799.6
n-Heptane	2077.6	2077.5
2-Methylhexane	2078.9	2079.2
3-Methylhexane	2078.6	2078.5
3-Ethylpentane	2078.3	2077.9
2,2-Dimethylpentane	2081.5	2081.9
2,3-Dimethylpentane	2079.8	2080.2
2,4-Dimethylpentane	2080.2	2080.9
3,3-Dimethylpentane	2081.2	2080.7
2,2,3-Trimethylbutane	2083.2	2081.5
2,2-Dimethylheptane	2642.5	2642.5
n-Decane	2919.1	2918.7
n-Pentadecane	4321.6	4320.7
n-Eicosane ($C_{20}H_{42}$)	5724.1	5722.7

Not only are differences among isomers exhibited by differences in heats of formation. Careful bond length measurements have also revealed small but definite differences that would be associated normally with differences in bond energy.[2]

Fortunately, this does not mean that atomization energies of organic compounds cannot be calculated with reasonable accuracy by the methods for polar covalence. It means merely that one must keep aware of small discrepancies and variations introduced by certain structural aspects of isomers. The following discussion will attempt, therefore, not to propose a new theory of organic chemistry but merely to indicate where organic chemists might look for new insights.

Because of the generally greater complexity of organic molecules, they tend to be susceptible to a much greater variety of chemical change wherein the options are not distinguished by large energy differences. In such situations seemingly minor differences assume much greater importance, and kinetics assumes a greater role than thermodynamics. Atoms tend to rearrange along the easiest paths where there are no clear advantages in product stability.

Any chemist can predict without hesitation the consequence of bringing chlorine into contact with sodium, but bringing chlorine into contact with butane is quite another story. The identity of the products will then depend much more on the availability of mechanisms than on their standard heats of formation. Contributing bond energies will still retain their fundamental importance, but bond dissociation energies become more interesting and informative.

CONTRIBUTING BOND ENERGIES

From the heat of combustion of methane may be determined the standard heat of formation, from which we can determine the average C—H bond energy of about 99 kpm. But it is interesting to know more about the bonding than merely the average energy. The conventional calculation of the polar covalent bond discloses that even in this very slightly polar bond, in which the ionic blending coefficient is only 0.03, the ionic contribution to the total energy is very substantial, about 9 kpm or 10% of the total.

For most ordinary organic compounds, it is possible to avoid most of the difficulty of nonconstant C—C and C—H bonds by assigning average values determined from a study of about 60 such derivatives. It is found that by assigning 98.7 kpm for each C—H bond except aromatic and tertiary and calculating all other bond energies in the usual way, very satisfactory representations of the total atomization energy can be obtained. Aromatic C—H is assigned 97.3 kpm, and tertiary C—H is calculated in the normal manner. The possibility of error increases with the number of separate bonds in the molecule, but the calculated heat of formation should not differ from the correct experimental value by more than about 1 kpm for every four bonds.

[2] D. R. Lide, *J. Chem. Phys.* **33**, 1514, 1519 (1960)

Indeed the successful application of bond energy calculations to complex organic molecules strongly supports the validity of the principles and methods applied. For example, a molecule of acetic acid contains seven different bonds of five different kinds. The sum of the calculated energies is 773, compared to the experimental value of 775 kpm. Table 11-3 summarizes the results of calculating

TABLE 11-3

Contributing Bond Energies and Atomization Energies of Representative Organic Compounds

Compound	Bond	$t_c E_c$	$t_i E_i$	E	Number of bonds	Total	Atomization energy (kpm) Calc.	Atomization energy (kpm) Exp.
CH_3Cl	C–H			98.7	3	296.1	380	377
	C–Cl	60.9	22.5	83.4	1	83.4		
CH_3COCH_3	C–H			98.7	6	592.2	935	938
	C–C	84.3	0	84.3	2	168.6		
	C=O″	114.3	60.2	174.5	1	174.5		
CH_3CH_2OH	C–H			98.7	5	493.5	770	771
	C–C	82.7	0	82.7	1	82.7		
	C–O′	46.1	34.8	80.9	1	80.9		
	O′–H	50.2	62.3	112.5	1	112.5		
$(CH_3CO)_2O$	C–H			98.7	6	592.2	1330	1325
	C–C	85.4	0	85.4	2	170.8		
	C–O″	72.1	38.0	110.1	2	220.2		
	C=O″	113.4	59.8	173.2	2	346.4		
C_6H_5CHO	C–H			97.3	5	486.5	1582	1581
	C–H	87.6	9.1	96.7	1	96.7		
	C–C	122.6	0	122.6	6	735.6		
	C–C	85.4	0	85.4	1	85.4		
	C=O″	116.1	61.2	177.3	1	177.3		
CH_3CH_2SH	C–H			98.7	5	493.5	737	733
	C–C	83.2	0	83.2	1	83.2		
	C–S	64.5	7.3	71.8	1	71.8		
	S–H	70.9	17.5	88.4	1	88.4		
$(CH_3)_3N$	C–H			98.7	9	888.3	1105	1100
	C–N	54.1	18.1	72.2	3	216.6		
$C_6H_5CH=CH_2$	C–H			98.7	3	296.1	1748	1753
	C–H			97.3	5	486.5		
	C–C	122.6	0	122.6	6	735.6		
	C–C	85.4	0	85.4	1	85.4		
	C=C	144.5	0	144.5	1	144.5		
$CH_3COOCH_2CH_3$	C–H			98.7	8	789.6	1319	1323
	C–C	83.4	0	83.4	2	166.8		
	C–O′	45.2	34.1	79.3	1	79.3		
	C–O″	72.1	38.0	110.1	1	110.1		
	C=O″	113.4	59.8	173.2	1	173.2		
C_6H_5Br	C–H			97.3	5	486.5	1293	1290
	C–C	122.6	0	122.6	6	735.6		
	C–Br	57.2	14.0	71.2	1	71.2		

atomization energies of eleven representative organic molecules averaging eleven bonds each. The average difference between calculated and experimental atomization energy is only about 3 kpm, which would also be the difference between heats of formation. Percentagewise, the average difference between calculated and experimental atomization energy is only 0.3%.

STANDARD BOND ENERGIES

Many textbooks provide tables of "standard" bond energies such that addition of the appropriate quantity for each bond in a molecule should provide the total atomization energy of the molecule. Or, subtracted from the total experimental atomization energy, they can provide an approximate evaluation of one bond previously unknown. These tables have been derived purely empirically. They are reasonably accurate in some applications but not in others.

The only contributing bond energies that can be determined from experimental data are "average bond energies," obtained by dividing the total atomization energy of a molecule in which all the bonds are alike by the number of bonds. It is then assumed that these average bond energies will be applicable to other compounds in which there may be more than one kind of bond and therefore no experimental method for determining their energy. For example, having found the C—H energy as the average for methane, one can then determine the C—Cl energy in CH_3Cl by subtracting the sum of three C—H bond energies from the total atomization energy as determined experimentally. In this manner extensive tables have been built. The only reason they work at all is that the energy of a given bond does not vary greatly from one environment to another unless the environmental change is rather drastic. Even though the partial charge on a given atom may change considerably, if it becomes more positive the other atom becomes simultaneously less negative so that the average of the two, or ionic blending coefficient t_i, remains fairly constant resulting in similar calculated bond energy.

It is easy to show that the "standard" energies do have a fundamental explanation. The best way is to show the similarity between calculated and empirical values. However, it is nearly as effective to calculate the polar covalent energy for an imaginary diatomic molecule having a single covalent bond and consisting of the appropriate elements. When this is done, it is found that the values obtained are very similar to those long known empirically. Evidence of this is provided in Table 11-4.

In summary, standard bond energies as commonly tabulated can be roughly useful but not dependable in the sense of a specific CBE calculation for a particular bond in a particular compound. However, the theoretical basis for CBE is evidently suitable for explaining the general order of magnitude of the empirical bond energy values.

TABLE 11-4
Standard Bond Energies—Empirical
and Calculated

Bond	Experimental	Calculated
C–H	99	98
N–H	93	93
P–H	76	76
O–H	111	113
S–H	82	87
C–C	83	83
C=C	146	144
C–O'	84	84
C=O"	192	192
C–S	65	71
C–Cl	78–81	84
C–Br	66–68	71
C–I	51–57	55
Si–Cl'''	86	94
N–F	64–65	67
P–Cl"	79	79
P–Br'	65	63
P–I'	51	47
O–F	44–45	48
O–Cl	48–52	49
S–Cl	60	66

MORE ABOUT CONTRIBUTING BOND ENERGY

From the viewpoint of understanding chemical reactions and the distribution of products in organic chemical changes as well as the chemical properties of various compositions and structures, it is clear that bond dissociation energies have much more to offer, directly, than contributing bond energies. The latter are essential and useful in accounting for standard heats of formation and therefore relative stabilities of compounds and the heats of reaction involved in the formation of certain products. However, they do not offer much help toward understanding or predicting the differences among bonds between the same atoms but in different environments. For example, the contributing bond energies of the C–Cl bond in benzyl chloride, methyl chloride, and chlorobenzene all lie in the range 83–87. Yet the bond dissociation energies are 70, 83, and 102 kpm. If breaking this bond is an essential part of any given chemical reaction, then the difference in dissociation energy will be far more significant than the similarity in contributing energy.

It is important, however, to recognize that the contributing bond energy is always a very important part of the bond dissociation energy and that the differences arise from differences in the reorganizational energies of the fragments or free radicals formed by the rupture of the bond. What we really want to know, in addition to the contributing bond energy which can be calculated so successfully, is the reorganizational energy. Chapter Ten describes how this energy has been evaluated for a number of common free radicals. We have here the opportunity to indicate by a few examples how these reorganizational energies may be applied in the interpretation of organic chemistry. Teachers of that vast subject will recognize the possibilities for extensions of these ideas.

A couple of additional comments may be worthwhile before the topic of contributing bond energies in organic compounds is left behind. One is that oxygen exhibits two types of single bond. Normally the O' energy of 33.5 kpm is involved in ethers, alcohols, etc. However, when the oxygen is bonded to a carbon atom that already possesses one oxygen atom, a carbonyl group $C=O$, then the O'' energy of 68.7 is applicable. This causes the bond energy to increase from around 80 to about 110 kpm. For example, in diethyl ether both bonds from carbon to oxygen have the energy of about 80 kpm each. However, in ethyl acetate the ether oxygen is attached to the ethyl group by an 80 kpm bond but to the acetate carbon by a bond of about 110 kpm. It is probable that carbon–nitrogen bonds to carbonyl carbon are similarly enhanced, although available data are inadequate for judging. If this were true, then amides and polyamides would exhibit stronger bonds between nitrogen and carbonyl carbon than the normal, fully weakened single C–N bond exhibits in amines. The means by which the lone pair weakening effect is thus reduced through carbonyl carbon is not understood, except that the same O'' single bond energy is of course applicable to the double $C=O$ bond on the same carbon atom.

A second point is that nowhere during the extensive study of organic molecules and their bond energies has there appeared any suggestion or suspicion of complications resulting from differences in carbon atom hybridization. There is no evidence from bond energy calculations that this is a factor to be considered. For example, the only difference in calculated bond energy between the C–Cl bond in methyl chloride and in phenyl chloride seems fully accounted for in terms of the shorter bond in the latter (which may or may not be the consequence of changes in the bonding orbitals). By the conventional methods the correct total atomization of each is obtained as the sum of the normal contributing bond energies, including normal single bonds for C–Cl in both molecules. The same is true for other halides.

SOME APPLICATIONS OF REORGANIZATIONAL ENERGIES

Relative Stabilities of Hydrocarbon Radicals

Isomeric organic compounds have long been recognized to exhibit distinct differences in the relative reactivities of their functional groups. It is found experimentally that such differences correspond to differences in bond dissociation energies. These can now be translated to reorganizational energies. The salutary effect of substituting carbon for hydrogen on methane has been pointed out in the preceding chapter. In detail, liberation of the methyl radical from a bond appears to destabilize the residual CH_3 by about 2 kpm, which therefore causes the breaking of the bond to require 2 additional kpm. Where one of the hydrogen atoms is replaced by methyl forming the ethyl radical, however, then breaking loose the ethyl radical stabilizes the residual C_2H_5 by about 1 kpm, which means that the bond breaking requirement is reduced by that amount. When still another methyl group replaces a methyl hydrogen forming the isopropyl radical, breaking loose this radical stabilizes the residual C_3H_7 by about 2 kpm, easing the breaking of the bond by that amount. Finally, substitution of the final hydrogen by methyl forming the t-butyl radical stabilizes the radical still more, reducing the energy for bond breaking by 5 kpm. In summary, the reorganizational energies are: CH_3, 2; C_2H_5, -1; $i-C_3H_7$, -2; and $t-C_4H_9$, -5 kpm.

Very small differences in reorganizational energies, and thus radical stabilities, have profound effects on the rate at which they may form, the most stable radical tending to form most easily. It is possible to relate these reorganizational energies, then, to the relative reactivities of hydrocarbons, alcohols, and halides with respect to free radical mechanisms: $t-C_4H_9 > i-C_3H_7 > C_2H_5 > CH_3$.

Furthermore, it is also found that the ease of ionization of the radical decreases in the same order. The values are: CH_3, 229; C_2H_5, 202; $i-C_3H_7$, 182; and $t-C_4H_9$, 171 kpm. This means that the stabilities of the carbonium ions formed by removing one electron from each also decrease in the order t-butyl . . . methyl. These differences are greater than between radical stabilities and thus have a large effect on the course of reactions involving carbonium ions.

One does not reject the contributing bond energies of bonds in isomers or otherwise similar compounds because they fail to reveal the differences that govern the reactions. One simply recognizes that the contributing bond energies are essentially correct but that the reorganizational energies of the radicals are therefore the principal influence on the course of reaction.

The great utility of the, at least approximately, valid assumption of constant reorganizational energy is that once this value is known for a given radical, **it applies to all possible chemical combinations of that radical** and does not need to be determined separately, as a bond dissociation energy, for each individual compound of that radical.

Alkyl, Vinyl, and Aryl halides

Organic halides are among the most important intermediates in organic synthesis. Therefore their relative reactivities are of great practical importance. Halogen attached to an unsaturated or aromatic carbon has long been recognized to be much less reactive than halogen attached to an alkyl carbon. Various explanations have been offered. One relates to the state of carbon hybridization, comparing sp^2 with sp^3. However, as stated earlier, the evidence for significant differences caused by such variations in hybridization is really not very convincing. Furthermore, an sp² hybrid bond, having greater "s character," is often alleged to involve more electronegative carbon, which would cause the bond to halogen to be of reduced ionicity and therefore weaker, not stronger. This is belied by the significantly shorter bond of halogen to unsaturated carbon. Perhaps the main argument assumes resonance involving partial double bond character of the carbon–halogen bond. This seems an unnecessary complication. We need only recognize that the reorganizational energies of the vinyl and phenyl radicals are relatively high, +12 and +15 kpm, whereas the reorganizational energies of the alkyl groups are low or negative. These differences alone are quite adequate to cause the bond dissociation energy for the halogen–carbon bond to be much lower for the alkyl halides. It follows that the halide reactivity should be greater for the alkyl derivatives.

Allyl Reactivity

In contrast to the vinyl compounds, in which halogen is attached directly to unsaturated carbon, the allyl grouping leaves a single bond in between. The result is that removal of a halogen atom, or other constituent, leaves the allyl radical with the possibility of having its double bond in either position. A general principle is that wherever such a situation occurs, the tendency is for the electrons to make maximum use of all possibilities (resonance) instead of adopting one to the exclusion of the others. Therefore in the allyl radical the two carbon-to-carbon bonds become equalized, neither being either single or double but both being intermediate. This is a more stable arrangement than one single and one double bond could be, in part because a 1.5 bond does not require so high a concentration of electrons within the bonding region. (For example, the energy of the 1.5 bond in benzene, 123 kpm, is greater than the average of a single bond, 83, and a double bond, 144, which is 114 kpm.) Consequently an allyl radical is stabilized, as it is found from bond dissociation energies, by about −10 kpm. Thus it would be about 22 kpm easier to break loose an allyl radical than a vinyl radical from their respective combinations with the same other radical.

Summary

The use of bond dissociation energies to explain a considerable number of important organic chemical phenomena can now be improved by using, instead, the reorganizational energies of free radicals to the extent that they have been determined. Those especially skilled and experienced in the art will doubtless recognize the value of this approach and visualize more clearly the breadth of its possible application.

ORGANOMETALLIC COMPOUNDS

It may be recalled that compounds of negative hydrogen have bond energies substantially lower than calculated for them assuming polar covalence. For such compounds the bond energy seems better represented as if the bond to hydrogen were nonpolar. It is assumed that the very highly polarizable negative hydrogen becomes sufficiently polarized to cancel, in effect, the bond ionicity. This reduction in bond energy appears to have no effect on the chemical characteristics of the negative hydrogen, as judged by the known chemical properties of the compounds in which it occurs.

Similarly it has been found[3] that the physical and chemical properties of organometallic compounds, in particular alkyl metal compounds, are also closely related to the negative charge on the alkyl group, which serves as functional group in these compounds. Again, similarly, it is found that bond energies in the compounds of negative alkyl cannot be accurately calculated by the usual method for polar covalence because the total atomization energies so obtained are much too high. These results can be improved somewhat by ignoring the ionicity of the metal–carbon bond and treating it as nonpolar, but, unlike the situation with hydrides, this is not at all a satisfactory treatment. The partial charges on alkyl groups in such compounds range from −1.00 in lithium alkyls but are generally much smaller than in the corresponding hydrides or, at any rate, appear to have much less effect, as might be expected.

Data have been collected for all the methyl and ethyl compounds of the chemical elements for which the thermochemistry has been studied. This provides for study a body of 40 compounds, 25 of which have negatively charged (usually only slightly negatively charged) alkyl groups and the other 15 of which have positively charged alkyl groups, being mainly compounds of nonmetals. The average experimental atomization energy for the 40 compounds is about 2300 kpm. The average calculated atomization energy is 46 kpm higher than the experimental value for the 25 compounds of negative alkyl. In contrast the average calculated atomization energy for the 15 compounds of positive alkyl differs from the experimental value by only 7 kpm, or about 0.3%.

Unfortunately, there is no present explanation for this puzzling characteristic of these simple organometallic compounds. Perhaps the solution to this problem will be realized when improved thermochemical data for organometallic compounds become available. But it may be appropriate to conclude this chapter on a note of uncertainty, to emphasize what is oftimes lost sight of in the enthusiasm for the moment. Unless some catastrophic succession of events brings scientific research to a complete standstill and holds it there permanently, then any book such as this must be classified as "progress report" with the normal hope and expectation that its value as a stimulus to further investigation and study may at least equal its value in providing understanding.

COHESIVE ENERGY

The bond energy methods described in this book are applicable directly only to the gaseous state of molecular compounds, and to nonmolecular solids. Their

---

[3] R. T. Sanderson, *J. Am. Chem. Soc.* 77, 4531 (1955).

utility can be extended where reliable experimental cohesive energies for the molecular condensed phases are available, or where reasonable estimates of such cohesive energies can be made. Two excellent sources of thermochemical information have become available.[4] Both give far more extensive compilations of heats of formation of organic compounds in both gaseous and condensed states than were available for the work described here. In addition, the book by Cox and Pilcher points out that energy differences between gaseous and condensed states at 25° are known experimentally for only about half the organic compounds for which thermochemical measurements have been made. They therefore review many previously published methods of estimation, several of which could be satisfactorily applied to obtain standard heats of formation of the liquid compounds from calculated gas-phase atomization energies. Of these, the simplest are the equations of Wadso,[5] who estimates ΔH_v at 25° as a simple linear function of the boiling point in °C. For slightly or nonassociated liquids he gives:

$$\Delta H_v = 5.0 + 0.041 t_b (°C) \text{ kpm} \tag{1}$$

For alcohols:

$$\Delta H_v = 6.0 + 0.055 t_b (°C) \text{ kpm} \tag{2}$$

These equations appear reliable within a few tenths of a kilocalorie for most liquids boiling below about 200°C, but the relationship is not truly linear over a wider temperature range and larger errors result for higher boiling compounds.

Recently, I have been engaged in a study of cohesive energies in both liquids and solids. This is a complex and difficult area but certain useful information has already resulted which can appropriately be included here. More than 720 liquids are being studied. They include about 120 inorganic or organometallic compounds and more than 600 organic compounds. The inorganic compounds include at least five examples of each type: binary hydrogen, binary oxygen, binary sulfur, binary fluorine, binary chlorine, binary bromine, metal alkyls, oxyhalides, and a wide variety of miscellaneous compounds. The organic compounds include 157 alkanes, 90 alkenes, 52 cycloalkanes, 47 alkylbenzenes, 54 halides, 42 mercaptans and sulfides, 30 nitrogen compounds, 32 alcohols, 22 esters, 18 ethers, 12 ketones, 7 aldehydes, and a large number of other functional and mixed-functional compounds. These compounds have boiling points that range from −56 to 380°C or over 436°, and have a cohesive energy that ranges from 3.7 to 26.1 kpm or varies by a factor of 7.

Nearly 50 compounds having OH groups and therefore capable of protonic bridging are conspicuous for having cohesive energies appreciably higher than

---

[4] J. D. Cox and G. Pilcher, "Thermochemistry of Organic and Organometallic Compounds." Academic Press, New York, 1970; D. R. Stull, E. F. Westrum, Jr., and G. C. Sinke, "The Chemical Thermodynamics of Organic Compounds." Wiley, New York, 1969.

[5] I. Wadso, *Acta Chem. Scand.* **20**, 544 (1966).

expected for their boiling points. All the rest appear to conform quite well to the empirical equation:

$$C_l = \frac{T_b}{68.3 - 0.0674\,T_b} \qquad (3)$$

C_l is the cohesive energy in kpm and T_b the absolute boiling point. This is not an exact equation, but certainly suitable for estimating values in the calculation of standard heats of formation of liquid compounds in general from gaseous atomization energies. For about 98% of these compounds, the difference between calculated and experimental cohesive energies is less than 2 kpm, and for 93%, it is less than 1 kpm; 83% have differences of ±0.5 kpm or lower, 66% being within ±0.3 kpm. Wherever compounds differing by more than 1 kpm can be compared with other compounds of similar type, they appear exceptional, suggesting that the differences may originate with experimental error. Except for the alcohols and other OH compounds mentioned above, no *consistent* class deviations from Eq. (3) have been observed.

The cohesive energies of alcohols can be estimated quite well by adding 3.3 kpm to the value calculated from Eq. (3). For phenols, 1.9 kpm should be added. There is unquestionably a small structural effect on cohesive energy that is not quantitatively accounted for by change in boiling point, but in general this is less than 1 kpm and was adequately averaged out in the derivation of Eq. (3).

As might well be expected, the experimental cohesive energies of compounds that are crystalline solids at 25° show a more widely scattered relationship to boiling point. A similar study of 135 solid compounds, 41 inorganic and 94 organic, has produced the following approximate equation resembling Eq. (3):

$$C_s = \frac{T_b}{37.14 - 0.0172\,T_b} \qquad (4)$$

Again, OH compounds are exceptional. Of the others, the average difference between calculated and experimental cohesive energies is about 1.6 kpm.

Obviously, intermolecular cohesive energies make an important contribution to heats of formation of both inorganic and organic liquids and solids. A fuller understanding of chemical bonding and reaction must therefore await further studies of cohesive energies as well as of the chemical bonds themselves.

TWELVE

Summary:

Application to the

Oxidation of Ethane

This book has attempted to cover a large area indeed, and with only limited success. It seems appropriate that the final chapter be devoted to a very detailed example of the application of the principal ideas and methods described in this book, as a kind of summary of the kind of analysis that can now be made and the kinds of insights that may be obtained by such analysis. I have chosen the oxidation of ethane as providing illustrations of most of the important points that should be made. This is not necessarily a realistic oxidation of ethane, but rather a stepwise analysis that leads from ethane and oxygen to carbon dioxide and water.

THE NATURE OF ETHANE

From the standard heats of formation of carbon dioxide and water and from the heat of combustion of ethane, it is easily possible to calculate that the standard heat of formation of ethane is -20.2 kpm. This means, of course, that on the average the bonds that hold ethane molecules together are stronger than those that hold the same numbers of carbon atoms together in graphite and hydrogen atoms together in H_2 molecules of gas at $25°C$.

Let us begin by calculating the energies of the individual bonds in the C_2H_6 molecule. Here we are troubled by small variations in the $C-H$ bond energy, but we can calculate it one way and evaluate it another. The electronegativity of carbon is

3.79 and that of hydrogen, 3.55. Therefore the electronegativity of CH_3 (the empirical formula of ethane) is the fourth root of the product, $3.79 \times 3.55 \times 3.55 \times 3.55$, which is 3.61. In joining to form ethane, carbon atoms have changed in electronegativity by $3.61 - 3.79$ or -0.18. The change that would correspond to acquisition of unit charge is 4.05 (or 2.08 $\times$ the square root of 3.79). Therefore the partial charge on carbon in the ethane molecule is $-0.18/4.05 = -0.04$. Similarly, the electronegativity of hydrogen increases, when it forms ethane, from 3.55 to 3.61, or by 0.06. Unit charge would have caused a change of 3.92. Therefore the partial charge on the hydrogen is $0.06/3.92 - 0.02$ (actually 0.015). These charges are hardly enough to cause significant changes in atomic radius, and we may take the nonpolar covalent radii as essentially unchanged in the ethane molecule. The ionic blending coefficient for the C—H bond is the average of the partial charges on carbon and hydrogen, or 0.03, leaving the covalent blending coefficient as 0.97.

Since we must correct the homonuclear single covalent bond energy of hydrogen for the presence of partial charge when the charge is positive, we multiply 104.2, the normal bond energy, by 0.97, = 101.1. The geometric mean of this and 83.2, the homonuclear single covalent bond energy of carbon, is 91.7. We then calculate the energy of the C—H bond:

$$t_c E_c = 0.97 \times 91.7 = 88.9$$

$$t_i E_i = 0.03 \times 332/1.09 = 9.1$$

This gives a total C—H bond energy of 98.0 kpm. Multiplication by 6 gives 588 kpm. The C—C energy is 83.2, giving a sum for the molecule of 671.2 kpm. This is not in perfect agreement with the experimental value of 675.4 kpm. It seems preferable, although the difference is really quite small, to take the average bond energy for C—H, determined as previously described for a number of representative organic molecules, as 98.7. This gives a calculated atomization energy of C_2H_6 of 675.4 kpm, fortuitously exactly that value determined by experiment. In the subsequent calculations, we shall use the standard value of 98.7 kpm rather than attempt individual calculations for the C—H bond.

To summarize the formation of the ethane molecule, two carbon atoms have combined with six hydrogen atoms to share valence electrons in slightly polar single covalent bonds. The carbon atoms have become slightly negative at the expense of the hydrogen atoms, which have become slightly positive, but the contraction of the hydrogen and the expansion of the carbon appear for practical purposes to cancel one another and have negligible effect on the bond length. However, even a very small degree of ionicity will have an appreciable effect in increasing the bond energy, and we find that about 10% of the total C—H bond energy results from the 3% ionicity as indicated by the ionic blending coefficient. If the C—H bonds were not polar at all, then their bond energy would be the geometric mean of 104.2 for hydrogen and 83.2 for carbon, or 93.1 kpm. Thus the ethane molecule would be about 40 kpm less stable if its bonds were completely nonpolar.

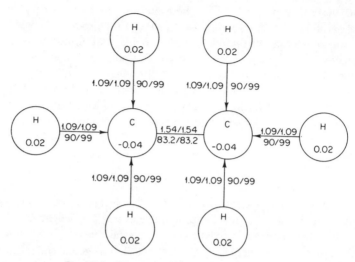

Fig. 12-1   Schematic representation of ethane.

Figure 12-1 is a schematic representation of the ethane molecule showing the partial charges, bond lengths, and bond energies according to the scheme outlined in Fig. 6-2, page 105.

Experimentally, it is found that dissociation of a C–H bond in ethane, leading to the separation of a hydrogen atom and an ethyl radical, requires 98 kpm. If we consider the calculated (standard) bond energy of 99 kpm, understanding that it represents the contributing bond energy for the C–H bond, and recall from Chapter Ten that the reorganizational energy of a free atom (such as H) is zero, and that the reorganizational energy of the ethyl radical is about −1 kpm, we can predict that the bond dissociation energy for splitting off one hydrogen atom should be about 98 kpm. Similarly, we can recognize the carbon–carbon bond energy in ethane as calculated to be the contributing bond energy, 83 kpm. Two moles of methyl radicals are liberated for each mole of C–C bonds dissociated. Therefore we can predict a bond dissociation energy of about 87 kpm. The experimental value is about 88 kpm.

OXIDATION STEP ONE: ETHANOL

The first step in the hypothetical oxidation of ethane would be the replacement of one of the hydrogen atoms by a hydroxyl group:

$$CH_3CH_3 + 0.5O_2 = CH_3CH_2OH$$

We know that the total bond energy in ethane is 675.4 kpm and that it would take half of 119.2, the dissociation energy of O_2, to produce each mole of oxygen atoms, or 59.6 kpm. The sum, 675.4 + 59.6, or 735.0 kcal, represents the relative stability of the system of ethane and oxygen molecules. To determine whether any advantage is to be derived from the combination to form ethanol, we need to examine the bonds in ethanol.

The electronegativity of oxygen is 5.21, so the electronegativity of the ethanol molecule is the geometric mean, or ninth root, of the product, 5.21 × 3.79 × 3.79 × 3.55 × 3.55 × 3.55 × 3.55 × 3.55 × 3.55, which is easily evaluated, using logarithms, as 3.76. Notice that addition of one oxygen atom to the ethane molecule increases its electronegativity from 3.61 to 3.76. This means that the carbon atoms must now be less negative and the hydrogen atoms more positive than in ethane, a natural consequence of the electron-withdrawing powers of the oxygen atom. The change in electronegativity of oxygen is −5.21 + 3.76 = −1.45, compared with the change of 4.75 corresponding to unit charge. Hence the partial charge on the oxygen in ethanol is −0.31. The electronegativity of carbon is only 0.03 higher than in the molecule of ethanol, which corresponds to a charge of −0.03/4.05 or −0.01 on carbon. The partial charge on hydrogen is now indicated by the gain in electronegativity of 0.21, which divided by 3.92 is 0.05. Thus the ionic blending coefficient for the C–H bond is still 0.03, the same as in ethane itself, for while the carbon atom was becoming less negative, the hydrogen atom was becoming more positive so that their average remained essentially unchanged.

The calculated C–H bond energy would be the same as in ethane; we shall use the value 98.7 kpm as suggested above. Five such bonds contribute a total of 493.5 kpm to the atomization energy of the ethanol.

From the partial charges of −0.01 on carbon and −0.31 on oxygen, we find the ionic blending coefficient to be the average (half the difference) or 0.15, leaving 0.85 for the covalent blending coefficient. The C–O bond length is experimentally determined as 1.43 Å, showing some shortening compared with the nonpolar covalent radius sum of 1.47 Å. The geometric mean of the homonuclear bond energy of carbon, 83.2, and the homonuclear bond energy of oxygen, 33.5, is 52.8 kpm. We may now calculate the contributing bond energy for the C–O single bond:

$$t_c E_c = \frac{0.85 \times 52.8 \times 1.47}{1.43} = 46.1 \text{ kpm}$$
$$t_i E_i = \frac{0.15 \times 332}{1.43} = 34.8 \text{ kpm}$$

The total C–O bond energy is then the sum of the two contributions, 80.9 kpm. The C–C energy is 83.2 × 1.54/1.55 = 82.7 kpm.

We now need only the O–H bond energy. The corrected homonuclear bond energy for hydrogen, with partial charge 0.05, is 0.95 × 104.2 or 99.0. The value for oxygen is again 33.5. The geometric mean for the O–H combination is 57.6 kpm. The ionic blending coefficient, taken from the partial charges on oxygen and hydrogen, is 0.18, leaving 0.82 for the covalent blending coefficient. The bond length experimentally determined is 0.96 Å, compared with the covalent radius sum of 1.02 Å. The polar bond energy is then calculated:

$$t_c E_c = \frac{0.82 \times 57.6 \times 1.02}{0.96} = 50.2 \text{ kpm}$$

$$t_i E_i = \frac{0.18 \times 332}{0.96} = 62.3 \text{ kpm}$$

The total is 112.5 kpm.

The total atomization energy of the ethanol molecule, which is the sum of the calculated contributing bond energies, is 769.6 kpm. This is in excellent agreement with the experimental value of 770.8 kpm.

It was calculated earlier that the energy required to atomize the requisite ethane and O_2 to produce ethanol is 735.0 kpm. The enthalpy difference is thus −35.8 kpm, there being ample advantage in stability of ethanol to justify converting the ethane and oxygen mixture. The overall reaction involves conversion of two moles of ethane and one mole of O_2 to two moles of ethanol, all as gases, so there is clearly a decrease in the number of moles of gas and a corresponding decrease in entropy. We would therefore expect the free energy for such a reaction, if it could be done, would be somewhat less than the enthalpy change but still very negative. In fact the free energy change for the reaction is about −32.3 kpm, as expected.

In Fig. 12-2 the ethanol molecule is represented. The energies given therein are of course the calculated contributing bond energies. We can also make some predictions about the dissociation energies of some of the bonds in ethanol, with

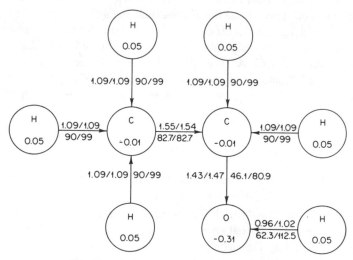

Fig. 12-2 Schematic representation of ethanol.

reference to Table 10-1. One point of rupture is of course the C–C bond. This has a contributing bond energy of 83 kpm. Its dissociation produces a methyl radical, for which the reorganizational energy is +2, and a $-CH_2OH$ radical, for which the reorganizational energy appears close to −2. We would therefore predict the same dissociation energy, or 83 kpm. The experimental value is 84 kpm. Or the dissociation could break a C–H bond on the hydroxyl carbon. The contributing bond energy here is about 99 kpm, and the liberated hydrogen atom would have zero reorganizational energy. The CH_3CHOH radical appears to have an E_R value of about −4, which suggests that the bond dissociation energy should be 95 kpm. The experimental value is only 90 kpm, the location of the error not being shown.

Another possible point of dissociation is the hydroxyl hydrogen. This O–H bond has a contributing bond energy of about 113 kpm. The E_R value for the ethoxy radical appears to be close to 1 kpm, which would give a calculated bond dissociation energy of 114 kpm. The experimental value is only 103, showing something seriously amiss at this point.

On the other hand, splitting off the hydroxyl radical would create a hydroxyl radical and an ethyl radical. The contributing bond energy for this C–O bond is about 81 kpm. The reorganizational energies for OH and C_2H_5 are 10 and −1, causing the bond dissociation energy to be $81 - 1 + 10 = 90$ kpm. The experimental value is 91, in good agreement.

Finally, we may observe that the plus 0.05 charge on hydrogen and the −0.31 charge on oxygen provide the requisites for significant protonic bridging among ethanol molecules as well as between ethanol and water. It should therefore not be surprising that ethanol is normally liquid and also miscible with water.

There is one consistent consequence of this simple method for assigning partial charges within a molecule that may well cause some doubt or confusion. One would certainly hesitate to claim that the hydrogen atoms attached to carbon in the ethanol molecule have the same degree of reactivity as the hydrogen attached to oxygen. Yet the method of calculating charges makes no distinction among atoms of the same kind, assigning the identical partial charge to each. Is this reasonable or not? Personally, I do not find this situation disturbing, because, apart from the partial charge on hydrogen, the environment of hydrogen is far different in its bond with oxygen. From the viewpoint of releasing a proton, it may seem that there is a choice between leaving behind an electron pair on oxygen or on carbon. It seems very probable that the carbon, being of the same electronegativity as the oxygen in the ethanol molecule, should be a far better donor of an electron pair to a proton—be far more basic than the oxygen. Hence it should be much easier to remove a proton from the oxygen.

Furthermore, in view of the mobility of the valence electrons, it is extremely difficult to understand why charges on like atoms would not become equalized. An ethyl group is surely a better reservoir of electrons than an atom of hydrogen, and an oxygen atom attached to both will surely gain most of its negative charge from the ethyl group. As indicated in our analysis here of the ethanol molecule, the oxygen only acquires 0.05 of its charge from the hydrogen atom attached to it, but the other −0.25 comes from the ethyl radical and, in particular, its five hydrogen atoms. To claim that the carbon atom holding the hydroxyl radical is "more positive" than the carbon atom of the methyl group is in effect to insist that the first carbon atom is more electronegative than the second. This is not a condition logically capable of persisting when the means of adjustment toward equalization are so readily at hand as in the carbon to carbon bond. In other words, the general flow of electrons is toward the oxygen, but the individual atoms of the same kind must bear the same partial charge as evidence of their equal electronegativity.

OXIDATION STEP TWO: THE SECOND HYDROXYL

The addition of an oxygen atom to a molecule of ethanol might occur, assuming the formation of a second hydroxyl group, at either carbon atom. Let us first consider the addition of the oxygen to the same carbon that already bears the hydroxyl group. Later we can consider the alternative and weigh the probabilities. The reaction we are considering is

$$CH_3CH_2OH + 0.5O_2 \rightarrow CH_3CH(OH)_2$$

Actually, the methods of calculation make no distinction between this product and ethylene glycol, $CH_2(OH)CH_2(OH)$, so far as contributing bond energies are concerned. First we find that the geometric mean electronegativity for the molecule is 3.88, determined in the usual manner. This corresponds to partial charges of 0.02 on carbon, 0.08 on hydrogen, and −0.28 on oxygen. Notice that the carbon is now positive, the hydrogen more positive than before, but the oxygen less negative. Naturally so, however, for now there are two oxygen atoms in the same molecule, both competing for the electrons of the other atoms.

Just as for the ethanol molecule, the C−H bond energy is 98.7, which multiplied by 4 gives 394.8 kpm. The C−C bond is slightly weakened by the positive charge on the carbon atoms, 83.2 × 0.98, but slightly shorter, 1.53 Å instead of 1.54 Å, and thereby strengthened. The value of 82.1 kpm is obtained. The C−O bond is exactly as in ethanol except that there are now two of them, for a total energy of 161.8 kpm. The O−H bond energy is somewhat different, owing to a different corrected homonuclear covalent bond energy for hydrogen as well as a slightly greater experimental bond length of 0.97:

$$t_cE_c = \frac{0.82 \times 56.7 \times 1.02}{0.97} = 48.9 \text{ kpm}$$

$$t_iE_i = \frac{0.18 \times 332}{0.97} = 61.6 \text{ kpm}$$

The total is 110.5 kpm, which for two bonds is 221.0 kpm. The grand total of the contributing bond energies so calculated is 859.7 kpm, which is a little lower than the experimental value of 867.5 kpm but still in fair agreement. Possibly internal protonic bridging occurs in ethylene glycol to increase the standard heat of formation experimentally determined. The above experimental atomization energy must be based on the glycol, of course, because the compound indicated as the product in our reaction does not exist stably. It loses water and becomes acetaldehyde:

$$CH_3 CH(OH)_2 \rightarrow CH_3 CHO + H_2O$$

We need to recognize two factors that contribute to this behavior. One is of course the proximity of the two hydroxyl radicals on the same carbon, which characteristically appears to allow an easy mechanism for the splitting out of a water molecule. The other is the enthalpy consideration. The total experimental atomization energy of acetaldehyde is 650.4 and that of water 221.6 kpm, making a total of 872.0 kpm. This is only about 5 kpm greater than for the dihydroxyethane, but it also involves an increase in entropy owing to the creation of a second mole of gas. Therefore an appreciable negative free energy is expected for the spontaneous splitting out of a water molecule from the dihydroxide.

Obviously it should be more difficult for water to split from a molecule of ethylene glycol, $CH_2 OHCH_2 OH$. The experimental atomization energy of the organic product, ethylene oxide, also is only 623 kpm, which added to that of water gives about 845 kpm. Here the splitting out of water is not favored either by proximity of hydroxyl groups or by enthalpy change. Let us continue then to consider principally the combination of oxygen with the same carbon atom of ethane.

Water

Since water is a product, we need to study it as well as the acetaldehyde molecule. The electronegativity of the molecule is the cube root of the electronegativity product, $5.21 \times 3.55 \times 3.55$, or 4.03. This means that the oxygen has gained partial negative charge at the expense of the hydrogen atoms. The partial charge on oxygen is the ratio of the actual change in electronegativity, $5.21 - 4.03 = 1.18$, to the change corresponding to unit charge on oxygen, or 4.75. Since the electronegativity of oxygen decreases, this corresponds to a negative sign of the charge, -0.25. The electronegativity of hydrogen is increased in water by $4.03 - 3.55 = 0.48$, which divided by 3.92 (the unit charge change) is 0.12 for the partial charge on each hydrogen atom. We therefore take 0.18 as the ionic blending coefficient. The bond length is observed to be 0.96 Å, compared to the covalent radius sum of 1.02 Å.

The hydrogen homonuclear bond energy must of course be corrected: $104.2 \times 0.88 = 91.5$. The geometric mean with 33.5 for oxygen is 55.4 kpm. We may now calculate the bond energy in water:

$$t_c E_c = \frac{0.82 \times 55.4 \times 1.02}{0.96} = 48.3 \text{ kpm}$$

$$t_i E_i = \frac{0.18 \times 332}{0.96} = 62.3 \text{ kpm}$$

The total, 110.6 kpm, is the calculated O–H bond energy in water. But there are two such bonds, causing the total atomization energy of water to be 221.4 kpm. The experimental value is 221.6 kpm.

A schematic representation of a water molecule is included in Fig. 8-2, page 120. Clearly each water molecule presents several possible sites for protonic bridging, in fact four, two for each hydrogen and two for each oxygen atom, by way of the lone pair electrons. We can account for the structure of water and ice in terms of this extensive protonic bridging. Since a coordination number of four is much smaller than the maximum possible for a water molecule, the rigid structure of ice represents a lower density than the more random liquid at the same temperature, and water exhibits the unusual property of expanding when it freezes.

It is interesting to note above that although the bonds in water are only 18% ionic, as determined by the charge average, they contribute more than half of the total energy through their ionic components.

Acetaldehyde

Now turning our attention to acetaldehyde, the product remaining after two oxygens have reacted with ethane and one water molecule has split out, we can calculate the electronegativity to be 3.82, a little lower than in the dihydroxy compound but still higher than the 3.76 calculated for ethanol. In the usual manner, we calculate partial charges of 0.01 on carbon, 0.07 on hydrogen, and −0.29 on oxygen. The customary practice is to assign 98.7 for the bond energy of a C–H bond on the methyl but calculate separately the bond energy of the tertiary C–H, which is 96.2. The carbon–carbon energy is slightly reduced by the small positive charge on each carbon atom, but the bond is only 1.50 instead of the covalent radius sum of 1.54, suggesting a slightly stronger bond. The C–C energy is then $83.2 \times 0.99 \times 1.54/1.50 = 84.8$ kpm.

The carbon to oxygen bond is now a double bond, of length 1.22 Å compared with 1.47 for the covalent radius sum. This means that the oxygen bond energy is the E'' energy of 68.7 kpm instead of the E' energy of only 33.5. Formation of the double bond half removes the "lone pair weakening effect." This applies, however, only to the covalent portion of the bond, whereas the multiplicity factor, 1.5, is applicable both to covalent and ionic contributions. The geometric mean of 68.7 and 83.2 is 75.6 kpm. The bond energy may now be calculated:

$$t_c E_c = \frac{0.85 \times 75.6 \times 1.5 \times 1.47}{1.22} = 116.1 \text{ kpm}$$

$$t_i E_i = \frac{0.15 \times 1.5 \times 332}{1.22} = 61.2 \text{ kpm}$$

The total is 177.3 kpm for the C=O bond.

The sum of all the calculated contributing bond energies in the acetaldehyde molecule is 656.9 kpm, compared to the experimental value of 650.4 kpm. The

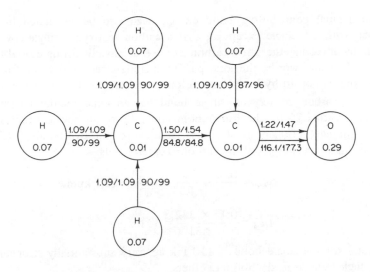

Fig. 12-3 Schematic representation of acetaldehyde.

difference of about 1% is probably to be considered reasonably satisfactory considering the six bonds of four different kinds in the molecule. Figure 12-3 gives a representation of the acetaldehyde molecule.

Although the above calculated contributing bond energies are reasonably correct, they are not the same as the bond dissociation energies. For example, breaking the carbon–carbon bond would liberate both a methyl radical and an aldehyde radical. The reorganizational energy of the former is +2 kpm, but the latter has a value of −15 kpm. The contributing bond energy was calculated to be 85 kpm, so the bond dissociation energy would be increased over this by 2 kpm and decreased by 15, a net decrease of 13 kpm to 72 kpm. The experimental value is about 70 kpm.

Similarly, to dissociate a hydrogen atom from the aldehyde group would involve the contributing bond energy of about 97, together with the reorganizational energy of the acetyl group, CH_3CO, which appears to be about −13. The calculated bond dissociation energy is then 84, which is in fair agreement with the experimental value of 87 kpm.

OXIDATION STEP THREE: ACETIC ACID

Addition of one more oxygen atom to a molecule of acetaldehyde would produce a molecule of acetic acid, without any further splitting out of water:

$$CH_3CHO + 0.5O_2 \rightarrow CH_3COOH$$

The electronegativity of the acetic acid molecule is 3.97, considerably higher than for the preceding oxidation step. This value corresponds to partial charges, calculated in the usual manner, of carbon 0.04, hydrogen 0.11, and oxygen −0.26.

The individual contributing bond energies can also be calculated in the usual manner, with the single exception that a carbon to oxygen single covalent bond profits by involving the same carbon atom that already forms a double bond to oxygen. In other words, the lone pair weakening effect is reduced for the single bond from carbon to hydroxyl as it is for the double bond from carbon to oxygen.

The carbon to oxygen single bond has an experimental length of 1.31, compared with 1.25 for the double bond and 1.47 for the covalent radius sum. The ionic blending coefficient is of course identical for the two bonds, 0.15. We may now calculate the energies of the carbon–oxygen bonds:

$$t_c E_c = \frac{0.85 \times 75.6 \times 1.47}{1.31} = 72.1 \text{ kpm}$$

$$t_i E_i = \frac{0.15 \times 332}{1.31} = 38.0 \text{ kpm}$$

The total for the single bond is 110.1 kpm, very substantially stronger than the usual single bond as in alcohols and ethers.

For the double bond, the length is less and the factor 1.5 applies:

$$t_c E_c = \frac{0.85 \times 75.6 \times 1.5 \times 1.47}{1.25} = 113.4 \text{ kpm}$$

$$t_i E_i = \frac{0.15 \times 1.5 \times 332}{1.25} = 59.8 \text{ kpm}$$

The total bond energy is 173.2 kpm.

The sum of the calculated contributing bond energies, using 98.7 for each hydrogen to carbon bond and calculating a C–C energy of $83.2 \times 0.96 \times 1.54/1.50$ = 82.0 kpm, is 661.4, up to the O–H bond.

The O–H bond has an experimental length of 0.95 Å. Correcting the homonuclear single covalent bond energy of hydrogen for the positive charge on hydrogen, we get 92.7. The geometric mean of this and 33.5 for the oxygen is 55.7 kpm. The energy contributions can now be calculated:

$$t_c E_c = \frac{0.82 \times 55.7 \times 1.02}{0.95} = 49.0 \text{ kpm}$$

$$t_i E_i = \frac{0.18 \times 332}{0.95} = 62.9 \text{ kpm}$$

The total O–H bond energy is thus 111.9 kpm. This adds to the sum above giving a total atomization energy of acetic acid, 773.3 kpm. The experimental value is known to be 774.9 kpm, in excellent agreement. Consider that there are five different kinds of bonds among the seven bonds in one molecule of acetic acid.

Figure 12-4 represents a molecule of acetic acid in the familiar manner. It will be recognized that the molecule is well equipped to form protonic bridges, which accounts for the liquid nature of acetic acid, its miscibility with water, and its formation of a dimer in the vapor state.

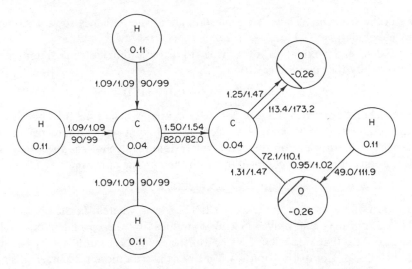

Fig. 12-4 Schematic representation of acetic acid.

The contributing bond energy that holds the hydroxyl radical to the remainder of the acetic acid molecule is about 80 kpm. Breaking that bond would become more difficult by 10 kpm because of the reorganizational energy of the hydroxyl radical, but less difficult by 13 kpm because of the exothermic reorganization of the acetyl radical. We would therefore predict a bond dissociation energy of $110 - 13 + 10 = 107$ kpm. This happens to be exactly the experimental value.

On the other hand, the reorganizational energy for the acetate radical appears to be about 6 kpm, which would make the bond dissociation energy for the acidic hydrogen, $113 + 6 = 119$ kpm. The experimental value of only 111 kpm discloses a discrepancy of unknown origin. The instability of the acetate radical itself has already been discussed in Chapter Ten.

OXIDATION COMPLETION: CARBON DIOXIDE AND WATER

The acetic acid molecule is quite resistant to oxidation. The process must require a number of intermediate steps, but let us merely consider the final products, carbon dioxide and water. The reaction would be

$$CH_3COOH + 2O_2 \rightarrow 2CO_2 + 2H_2O$$

The total atomization energy of the reactants would be $775 + 238 = 1013$ kpm. The total atomization energy of the products would be $2 \times 384 + 2 \times 222 = 768 + 444 = 1212$ kcal. The bonds in the products are thus about 200 kcal stronger than those in the reactants. There is no question about the tendency toward ultimately complete oxidation.

The overall oxidation of ethane to carbon dioxide and water can be represented by the chemical equation

$$2CH_3CH_3 + 7O_2 \rightarrow 4CO_2 + 6H_2O$$

The total bond energy in the reactants is 2185 kpm, whereas the total bond energy in the products is 2866 kpm, equivalent to 341 kpm heat evolved per mole of ethane burned. In view of the recent attention to alcohols as fuels to replace hydrocarbons, it is well to compare this with the burning of ethanol:

$$C_2H_5OH + 3O_2 = 2CO_2 + 3H_2O$$

The total bond energy in the reactants is 1128 kcal, whereas in the products it is 1434 kcal, meaning that only 306 kcal is produced by the burning of one mole of ethanol. Notice that this is less than 90% of the energy to be obtained by burning a mole of ethane. Comparison on a weight basis is even more discouraging, since a mole of ethanol is more than 50% heavier than a mole of ethane.

CONCLUSION

The familiar data of thermochemistry can of course supply us with most of the practical information that we may wish to obtain concerning applications of the kind illustrated here. They cannot, however, provide the understanding of bonding that this kind of analysis can so easily furnish. From a practical viewpoint, it seems rather futile to attempt to understand phenomena if the value of understanding does not justify the effort. This is inherently characteristic of quantum mechanical explanations of bonding. But the average practicing chemist, engaged in creating useful substances and controlling common chemical phenomena, can profit greatly by acquiring some degree of fundamental understanding of what he is doing and what it really means. The ideas, methods, and procedures of this book are intended to be helpful to the average chemist. This certainly also includes the average student of chemistry, whose limited capacity to handle abstract concepts should not deprive him of the opportunity to appreciate and enjoy the beautiful and seemingly simple cause-and-effect relationships between the properties of individual atoms and the properties of their chemical compounds. In this respect, the relatively narrow range of compounds used in this illustrative chapter should be recognized as only one out of millions of helpful examples that might be selected from almost any area of chemistry.

TABLE OF LOGARITHMS

N	0	1	2	3	4	5	6	7	8	9
1.0	.0000	.0043	.0086	.0128	.0170	.0212	.0253	.0294	.0334	.0374
1.1	.0414	.0453	.0492	.0531	.0569	.0607	.0645	.0682	.0719	.0755
1.2	.0792	.0828	.0864	.0899	.0934	.0969	.1004	.1038	.1072	.1106
1.3	.1139	.1173	.1206	.1239	.1271	.1303	.1335	.1367	.1399	.1430
1.4	.1461	.1492	.1523	.1553	.1584	.1614	.1644	.1673	.1703	.1732
1.5	.1761	.1790	.1818	.1847	.1875	.1903	.1931	.1959	.1987	.2014
1.6	.2041	.2068	.2095	.2122	.2148	.2175	.2201	.2227	.2253	.2279
1.7	.2304	.2330	.2355	.2380	.2405	.2430	.2455	.2480	.2504	.2529
1.8	.2553	.2577	.2601	.2625	.2648	.2672	.2695	.2718	.2742	.2765
1.9	.2788	.2810	.2833	.2856	.2878	.2900	.2923	.2945	.2967	.2989
2.0	.3010	.3032	.3054	.3075	.3096	.3118	.3139	.3160	.3181	.3201
2.1	.3222	.3243	.3263	.3284	.3304	.3324	.3345	.3365	.3385	.3404
2.2	.3424	.3444	.3464	.3483	.3502	.3522	.3541	.3560	.3579	.3598
2.3	.3617	.3636	.3655	.3674	.3692	.3711	.3729	.3747	.3766	.3784
2.4	.3802	.3820	.3838	.3856	.3874	.3892	.3909	.3927	.3945	.3962
2.5	.3979	.3997	.4014	.4031	.4048	.4065	.4082	.4099	.4116	.4133
2.6	.4150	.4166	.4183	.4200	.4216	.4232	.4249	.4265	.4281	.4298
2.7	.4314	.4330	.4346	.4362	.4378	.4393	.4409	.4425	.4440	.4456
2.8	.4472	.4487	.4502	.4518	.4533	.4548	.4564	.4579	.4594	.4609
2.9	.4624	.4639	.4654	.4669	.4683	.4698	.4713	.4728	.4742	.4757
3.0	.4771	.4786	.4800	.4814	.4829	.4843	.4857	.4871	.4886	.4900
3.1	.4914	.4928	.4942	.4955	.4969	.4983	.4997	.5011	.5024	.5038
3.2	.5051	.5065	.5079	.5092	.5105	.5119	.5132	.5145	.5159	.5172
3.3	.5185	.5198	.5211	.5224	.5237	.5250	.5263	.5276	.5289	.5302
3.4	.5315	.5328	.5340	.5353	.5366	.5378	.5391	.5403	.5416	.5428
3.5	.5441	.5453	.5465	.5478	.5490	.5502	.5514	.5527	.5539	.5551
3.6	.5563	.5575	.5587	.5599	.5611	.5623	.5635	.5647	.5658	.5670
3.7	.5682	.5694	.5705	.5717	.5729	.5740	.5752	.5763	.5775	.5786
3.8	.5798	.5809	.5821	.5832	.5843	.5855	.5866	.5877	.5888	.5899
3.9	.5911	.5922	.5933	.5944	.5955	.5966	.5977	.5988	.5999	.6010
4.0	.6021	.6031	.6042	.6053	.6064	.6075	.6085	.6096	.6107	.6117
4.1	.6128	.6138	.6149	.6160	.6170	.6180	.6191	.6201	.6212	.6222
4.2	.6232	.6243	.6253	.6263	.6274	.6284	.6294	.6304	.6314	.6325
4.3	.6335	.6345	.6355	.6365	.6375	.6385	.6395	.6405	.6415	.6425
4.4	.6435	.6444	.6454	.6464	.6474	.6484	.6493	.6503	.6513	.6522
4.5	.6532	.6542	.6551	.6561	.6571	.6580	.6590	.6599	.6609	.6618
4.6	.6628	.6637	.6646	.6656	.6665	.6675	.6684	.6693	.6702	.6712
4.7	.6721	.6730	.6739	.6749	.6758	.6767	.6776	.6785	.6794	.6803
4.8	.6812	.6821	.6830	.6839	.6848	.6857	.6866	.6875	.6884	.6893
4.9	.6902	.6911	.6920	.6928	.6937	.6946	.6955	.6964	.6972	.6981
5.0	.6990	.6998	.7007	.7016	.7024	.7033	.7042	.7050	.7059	.7067
5.1	.7076	.7084	.7093	.7101	.7110	.7118	.7126	.7135	.7143	.7152
5.2	.7160	.7168	.7177	.7185	.7193	.7202	.7210	.7218	.7226	.7235
5.3	.7243	.7251	.7259	.7267	.7275	.7284	.7292	.7300	.7308	.7316
5.4	.7324	.7332	.7340	.7348	.7356	.7364	.7372	.7380	.7388	.7396
N	**0**	**1**	**2**	**3**	**4**	**5**	**6**	**7**	**8**	**9**

Index

Abbreviations: AE, atomization energy; BDE, bond dissociation energy; CBE, contributing bond energy; RE, reorganizational energy; f, figure; t, table.

A

Acetaldehyde
AE of, 201
BDE in, 161t, 201
from ethanol exidation, 198, 200
representation of, 201f
Acetamide, BDE in, 162t
Acetate radical
decomposition of, 172
RE of, 168t
Acetic acid
AE of, 183, 203
bond dissociation energy in, 162t, 169, 172, 203
protonic bridging in, 115t
representation of, 203f
Acetic anhydride, CBE's, 183t
Acetone
BDE in, 164t
CBE's in, 183t
Acetyl bromide, BDE in, 164t

Acetyl chloride, BDE in, 164t
Acetyl iodide, BDE in, 164t
Acetyl radical, RE of, 162t
Acid, aqueous, 118, 138
Acid ionization, 118
Acid RE's, 176t
Advantages of new approach, 24, 95, 98, 204
Alcohols, bond lengths in, 93t
Aldehyde radical, RE of, 161t
Alkane bond energies, empirical, 181t
Alkyl compounds, bond length in, 93t
Allyl alcohol, BDE in, 163t
Allyl aldehyde, BDE in, 163t
Allyl amine, BDE in, 163t
Allylbenzene, BDE in, 163t
Allyl bromide, BDE in, 163t
Allyl chloride, BDE in, 163t
Allyl iodide, BDE in, 163t
Allyl reactivity, 188
Aluminum, selected properties of, 57, 67t
Aluminum arsenide, AE of, 109t

Aluminum borohydride, representation of, 118f
Aluminum bromide, AE of, 151t
Aluminum chlorides, gaseous, AE of, 150t
Aluminum fluoride, gaseous, AE of, 104t, 147t
Aluminum hydride, AE of, 113t
Aluminum iodide, AE of, 152t
Aluminum monoxide, gaseous, AE of, 128t
Aluminum nitride, AE of, 109t
Aluminum phosphide, AE of, 109t
Aluminum triethyl, RE of, 176t
Aluminum trimethyl, RE of, 176t
Amines, bond length in, 93t
Amino radical, RE of, 169t
Ammonia
  AE of, 103t, 113t
  BDE in, 169t
  protonic bridging in, 115t
Aniline, BDE in, 169t
Anions, nature of, 106
Anisole, BDE in, 167t
Antimony, selected properties of, 63, 67t
Antimony (III) chloride, gaseous, AE of, 150t
Arsenic, selected properties of, 61, 67t
Arsenic (III) bromide, gaseous, AE of, 151t
Arsenic (III) chloride, gaseous, AE of, 150t
Arsenic fluorides, AE of, 148t
Arsenic (III) iodide, AE of, 152t
Arsine, AE of, 113t
Aryl halides, 187
Atom
  energy of, 96
  nuclear model of, 26
Atomic charge, 77
Atomic compactness, 41
Atomic energy, electronic, 96
Atomic properties, selected values of, 52, 67t
Atomic radius, 3, 79
  in compounds, 79
Atomic structure, 26
Atomization energies, representative, 103t

B

Barium, selected properties of, 64, 67t
Barium bromide, solid, AE of, 154t
Barium chloride
  gaseous, AE of, 154t
  solid, AE of, 104t

Barium fluoride, solid, AE of, 153t
Barium fluorides, gaseous, AE of, 147t
Barium iodide, AE of, 155t
Barium oxide
  gaseous, AE of, 128t
  solid, AE of, 135t
Benzaldehyde
  BDE in, 161t
  CBE's in, 183t
Benzene, BDE in, 170t
Benzoic acid, BDE in, 161t
Benzoyl bromide, BDE in, 161t
Benzoyl chloride, BDE in, 161t
Benzoyl radical, RE of, 161t
Benzyl alcohol, BDE in, 162t
Benzyl amine, BDE in, 162t
Benzyl bromide, BDE in, 162t
Benzyl chloride, BDE in, 162t
Benzyl iodide, BDE in, 162t
Benzyl mercaptan, BDE in, 162t
Benzyl methyl sulfone, BDE in, 162t
Benzyl radical, RE of, 162t
Beryllium, selected properties of, 54, 67t
Beryllium bromide, solid, AE of, 154t
Beryllium bromides, gaseous, AE of, 151t
Beryllium chloride, solid, AE of, 154t
Beryllium chlorides, gaseous, AE of, 150t
Beryllium fluoride, solid, AE of, 104t, 153t
Beryllium fluorides, gaseous, AE of, 147t
Beryllium hydride, AE of, 113t
Beryllium iodide, solid, AE of, 155t
Beryllium iodides, gaseous, AE of, 152t
Beryllium oxide
  solid, AE of, 103t, 135t
  gaseous, AE of, 128t
Bifluoride ion, 116
Bismuth, selected properties of, 65, 67t
Bismuth tribromide, AE of, 151t
Bismuth trichloride, AE of, 150t
Bismuth monofluoride, AE of, 148t
Bismuth monoxide, AE of, 128t
Bismuth triiodide, AE of, 152t
Blending coefficients, 11, 101, 103
Bond
  covalent, 10
  ionic, 99
  metallic, 5
Bond dissociation, 156
Bond dissociation energy (BDE), 172
Bond energy
  calculation of by quantum mechanics, 11, 96
  contributing (CBE), 156

effect of ionicity on, 101, 102f
homonuclear, 69
  periodicity of, 43,
  relation to rS, 43, 44f,t
  weakened and unweakened, 18, 49t, 50f
  weakening of, 18, 45
significance of, 95
standard, 184, 185t
Bond lengths
  in alkyl compounds, 93t
  in binary compounds, 83t
  in binary hydrogen compounds, 91t
  in bromides, gaseous, 86t
  in bromides, solid, 89t
  in chalcides, solid, 92t
  in chlorides, gaseous, 85t
  in chlorides, solid, 88t
  correction for polarity, 99
  in fluorides, gaseous, 84t
  in fluorides, solid, 88t
  in halides, gaseous, of 18-shell elements,
    90t
  in iodides, gaseous, 87t
  in iodides, solid, 89t
  in more complex compounds, 93t, 94
  summary of calculation for binary com-
    pounds, 83t
Bond multiplicity, 46
Bond polarity, 22, 77
  effect on bond strength, 13, 101
Bond type, diagnosis of, in oxides, 125
Bond weakening, 18
  reduction of, 48
Bonding, representation of, 105f
Borazine, representation of, 120f
Boric acids
  AE of, 139t
  difluoro, AE of, 139t
  fluoro, AE of, 139t
Born–Mayer equation, 14, 100
Boron, selected properties of, 54
Boron bromides, AE of, 151t
Boron chlorides, AE of, 150t
Boron fluorides, AE of, 147t
Boron hydrides, 119
  AE of, 113t
Boron iodides, AE of, 152t
Boron oxides, AE of, 104t, 128t, 129
Boron trifluoride
  AE of, 104t
  RE of, 176t
Boron trimethyl, RE of, 176t
Boroxy radical, RE of, 173t

Bromides, gaseous
  calculation of bond lengths in, 87t
  AE of, 151t
Bromides, solid
  AE of, 154t
  calculation of bond length in, 89t
Bromine, selected properties of, 62, 67t
Bromine chloride, AE of, 150t
Bromine fluorides, AE of, 148t
Bromine monoxide, AE of, 128t
2,3-Butadiene, BDE in, 169t
n-Butane
  AE of, 181t
  BDE in, 165t, 166t
Iso-butane
  AE of, 181t
  BDE in, 163t
n-Butanol
  BDE in, 165t
  bond length in, 93t
t-Butanol, BDE in, 163t
1-Butene, BDE in, 163t, 169t
t-Butylamine, BDE in, 163t
n-Butylbenzene, BDE in, 162t
Iso-Butylbenzene, BDE in, 163t
t-Butylbenzene, BDE in, 162t
t-Butyl-benzyl, BDE in, 162t
t-Butylbromide, BDE in, 163t
t-Butylchloride, BDE in, 163t
t-Butyliodide, BDE in, 163t
t-Butyl mercaptan, BDE in, 163t
t-Butyl radical, RE of, 163t
n-Butyraldehyde, BDE in, 163t
Iso-Butyraldehyde, BDE in, 161t
n-Butyrate radical, RE of, 173t

C

Cadmium, selected properties of, 62, 67t
Cadmium chloride, AE of, 154t
Cadmium fluoride, AE of, 153t
Cadmium oxide, AE of, 109t, 135t
Cadmium selenide, AE of, 109t
Cadmium sulfide, AE of, 109t
Cadmium telluride, AE of, 109t
Calcium, selected properties of, 59, 67t
Calcium bromide, solid, AE of, 104t, 154t
Calcium chloride, solid, AE of, 154t
Calcium fluoride, solid, AE of, 104t, 153t
Calcium fluorides, gaseous, AE of, 147t
Calcium iodide, solid, AE of, 155t

Calcium monochloride, gaseous, AE of, 150t
Calcium oxide
  gaseous, AE of, 128t
  solid, AE of, 104t, 135t
Carbon, 16
  burning of, 16
  formation as graphite not diamond, 17, 74
  selected properties of, 54, 67t
Carbon chlorides, gaseous, AE of, 150t
Carbon dioxide
  AE of, 137f
  bond energy in, 19, 22, 128t
  BDE in, 23, 161t
  from ethane, 203
  polymer of, 21, 135t, 137f
  representation, polymer and monomer,
    137f
  why different from silica, 137
Carbon disulfide
  AE of, 103t
  BDE in, 161t
  monomeric nature of, 149
Carbon fluorides, gaseous
  AE of, 147t
  bonding in, 149
Carbon monoxide
  AE of, 19, 128t
  dimer of, 130
Carbon tetrachloride
  AE of, 150t
  crowding in, 149
Carbon–hydrogen bonds, 179, 182
Carbon–oxygen bonds, 140, 186
Carbonic acid, AE of, 139t
Carbonium ions, 187
Carbonyl sulfide, BDE in, 161t
Carbonyl radical, RE of, 161t
Carboxyl radical, RE of, 171
Cations, nature of, 106
Cesium, selected properties of, 64, 67t
Cesium bromide
  gaseous, AE of, 151t
  solid, AE of, 104t
Cesium chloride
  gaseous, AE of, 150t
  solid, AE of 154t
Cesium fluoride
  gaseous, AE of, 147t
  solid, AE of, 153t
Cesium hydride, AE of, 113t
Cesium iodide
  gaseous, AE of, 152t
  solid, AE of, 155t

Cesium oxide, solid, AE of, 135t
Chalcides, solid, calculated bond lengths in,
    92t
Chalcogens, physical state of, 71
Charge, see Partial charge
Chemical theory, advantages of, new ap-
    proach to, 24, 95, 98, 204
Chlorides, binary
  periodicity of charge in, 144t, 145f
  gaseous, AE of, 150t
  bond lengths in, 85t
  periodicity of bond energies in, 153f
Chlorides, solid
  AE of, 154t
  bond lengths in, 88t
Chlorine, selected properties of, 59, 67t
Chlorine fluorides, AE of, 148t
Chlorine monoxide, AE of, 128t
Chlorine trifluoride, AE of, 103t
Cohesive energy, 189
Contributing bond energy (CBE), 23, 182
Coordinate covalent model of solids, 14
Coordinated polymeric model, 15, 107
  advantages of, 108
Coordination in molecular addition com-
    pounds, 175
Copper, selected properties of, 66, 67t
Covalence, polar, representation of, 105f
Covalent blending coefficient, 11, 101, 103
Covalent energy, nonpolar, calculation of,
    14, 99, 100t
Cumene, BDE in, 165t
Cyclobutyl radical, RE of, 173t
Cyclohexyl radical, RE of, 173t
Cyclopentyl radical, RE of, 173t
Cyclopropyl radical, RE of, 173t

                              D

n-Decane, AE of, 181t
Decane isomers, 179
Diamond, atomization of, 17
Dibenzyl, BDE in, 162t
Diborane
  AE of, 113t
  representation of, 118f
Dichloroaluminum radical, RE of, 173t
Diethylamine, calculated bond length in,
    93t
Diethylether
  BDE in, 167t
  bond length in, 93t

Diethylperoxide, BDE in, 167t
Difluoroaluminum radical, RE of, 173t
Difluoroboric acid, AE of, 139t
Difluoronitrogen radical, RE of, 173t
Dimethylamine, calculated bond length in, 93t
2,2-Dimethylbutane
  BDE in, 163t
  AE of, 181t
2,3-Dimethylbutane, AE of, 181t
3,3-Dimethylbutene-1, BDE in, 163t
Dimethylether, BDE in, 167t
2,2-Dimethylheptane, AE of, 181t
2,3-Dimethylheptane, AE of, 181t
2,4-Dimethylheptane, AE of, 181t
3,3-Dimethylheptane, AE of, 181t
2,2-Dimethylpentane
  AE of, 181t
  BDE in, 163t
Dimethylperoxide, BDE in, 167t
2,2-Dimethylpropane, BDE in, 163t
1,2-Dimethylpropanol, BDE in, 164t
2,2-Dimethylpropanol, BDE in, 163t
Dimethylsulfide, BDE in, 166t
Dimethylsulfone, BDE in, 162t
Dinitrogen tetroxide, BDE in, 161t
Diphenyl, BDE in, 170t
1,2-Diphenylethane, BDE in, 162t
Diphenylketone, BDE in, 161t
Diphenylmethane, BDE in, 162t

E

Effective nuclear charge, 2, 33
  elimination of, 43
n-Eicosane, AE of, 181t
Electron affinity, 3
Electron vacancies, 31, 38
Electronegativity, 3, 5
  from atomic compactness, 39, 41t
  change with charge, 77
  Pauling, 36
  periodicity of, 42f
  principle of equalization of, 9, 75
  significance of, 50, 51
  table of, 41t
Electronic configuration, 26
  from ionization energies, 28
Electronic density, 40
Electronic energy of atom, calculation of, 96

Electrostatic model of solids, atomization energy from, 109
Energy
  cohesive, 189
  covalent, parameters for, 100t
  ionic, as substitute for part of covalent energy, 37
  polar covalent, calculation of, 97
  reorganizational, of free radicals, 157
  total atomic, 96
Energy profile of atom, 29
Entropy, 12, 15
ESCA, 79
Ethane
  AE of, 181t
  BDE in, 168t
  oxidation of, analysis of, 192
  representation of, 194f
Ethanol
  BDE in, 197
  bond length in, 93t
  CBE's in, 183t, 194
  oxidation of, 198
  protonic bridging in, 197
  radical, RE of, 167t
  representation of, 196f
Ethers, bond length in, 93t
Ethoxy radical, RE of, 167t
Ethyl acetate
  BDE in, 162t
  CBE's, 183t
Ethylamine
  BDE in, 166t
  bond length in, 93t
Ethylbenzene, BDE in, 162t
Ethylbenzoate, BDE in, 161t
Ethyl bromide
  bond length in, 93t
  BDE in, 166t
Ethyl chloride
  BDE in, 166t
  bond length in, 93t
Ethylene, BDE in, 169t
Ethylene glycol, bond length in, 93t
Ethylene oxide, AE of, 199
Ethyl iodide
  BDE in, 166t
  bond length in, 93t
Ethyl mercaptan
  BDE in, 166t
  bond length in, 93t
  CBE's of, 183t
3-Ethylpentane, AE of, 181t

Ethyl peroxide, BDE in, 167t
Ethyl phenyl ether, BDE in, 167t
Ethyl radical, RE of, 166t

F

Fluorides
  gaseous
  bond lengths in, 84t
    AE of, 147t
  solid
    AE of, 153t
    bond lengths in, 88t
Fluorine, selected properties of, 56, 67t
Fluoroxy radical, RE of, 173t
Formaldehyde, BDE in, 161t
Formic acid
  BDE in, 161t
  protonic bridging in, 115t
Free energy, 12, 15, 20
Functional group, 178

G

Gallium, selected properties of, 60, 67t
Gallium antimonide, AE of, 109t
Gallium arsenide, AE of, 109t
Gallium chlorides, gaseous, AE of, 150t
Gallium hydride, AE of, 113t
Gallium monobromide, gaseous, AE of, 151t
Gallium nitride, AE of, 109t
Gallium phosphide, AE of, 109t
Gallium triethyl, RE of, 176t
Gallium trimethyl, RE of, 176t
Germane, AE of, 113t
Germanium, selected properties of, 60, 67t
Germanium bromides, gaseous, AE of, 151t
Germanium chlorides, gaseous, AE of, 150t
Germanium dioxide, AE of, 135t
Germanium fluorides, gaseous, AE of, 147t
Germanium hydride, AE of, 113t
Germanium iodides, AE of, 152t
Germanium monoxide, gaseous, AE of, 128t
Germanyl radical, RE of, 173t
Graphite, 17, 74

H

Half bonds, 152
Halide chemistry, 142

Halides of 18-shell elements, bond lengths in, 90t
Halides
  chemical properties of, 143
  gaseous, bond energies in, 146
  hydrolysis of, 143
  organic, bond lengths in, 93t
  oxidizing power of, 143
  periodicity of, 144
  physical properties, 142
  solid, AE of, 153
Halogens
  bond dissociation energies of, 45, 49
  physical states of, 71
Helium family, 69
n-Heptane, AE of, 181t
n-Hexane
  AE of, 181t
  BDE in, 168t
Hybridization, 188
Hydrazine
  AE of, 113t
  BDE in, 169t
Hydrides
  complex, 117
  diatomic, 114
  solid, 119
Hydridic bridging, 116
Hydrocarbon radicals
  relative stability of, 186
  RE of, 187
Hydrogen, 111
  bond energy correction for, 112
  bonding, see Protonic bridging, Hydridic bridging
  charge on, periodicity of, 116, 121f
  covalent radius of, 53, 111
  electronegativity of, 112
  partial charge on, relation to chemistry, 117
  selected properties of, 53, 67t
Hydrogen bromide, AE of, 113t, 151t
Hydrogen chemistry, periodicity of, 116
Hydrogen chloride, AE of, 103t, 113t, 150t
Hydrogen compounds
  bond lengths in, 91t
  classification of, 121
  partial charge and properties of, 122t
Hydrogen fluoride
  AE of, 104t, 113t, 147t
  protonic bridging in, 115t
Hydrogen iodide, AE of, 113t, 152t
Hydrogen peroxide

AE of, 113t, 128t
BDE in, 169t
representation of, 129f
Hydrogen selenide, AE of, 113t
Hydrogen sulfide
AE of, 113t
BDE in, 167t
Hydrogen telluride, AE of, 113t
Hydrosulfide radical
AE of, 113t
RE of, 167t
Hydroxides, 139t
Hydroxyethyl radical, RE of, 164t
Hydroxyl radical
AE of, 113t, 128t
RE of, 160, 164t
Hydroxymethyl radical, RE of, 164t
Hypobromous acid, BDE in, 169t
Hypochlorite radical, RE of, 173t
Hypochlorous acid, BDE in, 169t
Hypoiodous acid, BDE in, 169t

I

Indium, selected properties of, 62, 67t
Indium antimonide, AE of, 109t
Indium arsenide, AE of, 109t
Indium bromides, AE of, 151t
Indium chlorides, AE of, 150t
Indium hydride, AE of, 113t
Indium iodides, AE of, 152t
Indium nitride, AE of, 109t
Indium phosphide, AE of, 109t
Indium trimethyl, RE of, 176t
Inert elements, 69, 106
Interhalogens, 152
Iodides
gaseous
AE of, 152t
bond lengths in, 87t
solid
AE of, 155t
bond lengths in solid, 89t
Iodine, selected properties of, 63, 67t
Iodine bromide, AE of, 151t
Iodine chloride, AE of, 150t
Iodine fluorides, AE of, 148t
Ionic blending coefficient, 11, 101
calculation of from bond energy, 103
calculation of from charges, 11, 101
Ionic energy, calculation of, 13, 37, 99
Ionic model, objections to, 105

Ionicity, effect of, on bond energy, 101, 102f
Ionization energy, 3
Ions in solid, behavior of, 106

K

Krypton, energy profile of, 29

L

Lead, selected properties of, 65, 67t
Lead bromides, AE of, 151t, 154t
Lead chlorides, AE of, 103t, 150t, 154t
Lead fluorides, AE of, 104t, 147t, 153t
Lead iodides, AE of, 152t, 155t
Lead (II) oxide, gaseous, AE of, 128a
Leveling effect, 119
Lithium, selected properties of, 53, 67t
Lithium bromide
gaseous, AE of, 151t
solid, AE of, 154t
Lithium chloride
gaseous, AE of, 150t
solid, AE of, 154
Lithium fluoride
gaseous, AE of, 147t
solid, AE of, 153t
Lithium hydride, AE of, 113t
Lithium hydroxide, gaseous, AE of, 139t
Lithium iodide
gaseous, AE of, 152t
solid, AE of, 155t
Lithium oxide, solid, AE of, 135t
Lone pair bond weakening, 18, 21, 45, 49t
reduction of, 48

M

M5 elements, physical state of, 73
M8 elements, physical state of, 69
Madelung constant, 14
Magnesium, selected properties of, 56, 67t
Magnesium bromide, solid, AE of, 154t
Magnesium bromides, gaseous, AE of, 151t
Magnesium chloride, solid, AE of, 154t
Magnesium chlorides, gaseous, AE of, 150t
Magnesium fluoride, solid, AE of, 153t
Magnesium fluorides, gaseous, AE of, 147t
Magnesium iodide, solid, AE of, 155t
Magnesium oxide

gaseous, AE of, 128t
solid, AE of, 135t, 136f
Manganese, selected properties of, 66, 67t
Manganese (II) bromide, AE of, 103t
Mercury, selected properties of, 64, 67t
Metaborate radical, RE of, 173t
Metallic bonding, 5
Methane
 AE of, 103t, 112t
 BDE in, 168t
Methanol
 BDE in, 167t
 bond length in, 93t
 protonic bridging in, 115t
 radical, RE of, 167t
Methoxy radical, RE of, 167t
Methyl acetate, BDE in, 162t
Methyl allyl ether, BDE in, 163t
Methyl allyl sulfone, BDE in, 162t
Methylamine
 BDE in, 168t
 bond length in, 93t
Methyl aniline radical, RE of, 173t
Methylbenzylether, BDE in, 162t
Methylbenzylketone, BDE in, 162t
Methylbenzylsulfide, BDE in, 162t
Methylbenzylsulfone, BDE in, 162t
Methyl bromide
 BDE in, 168t
 bond length in, 93t
1-Methylbutanol, BDE in, 164t
2-Methylbutene-1, BDE in, 163t
Methyl t-butyl sulfide, BDE in, 163t
Methyl t-butyl sulfone, BDE in, 162t
Methyl chloride
 BDE in, 168t
 bond length, 93t
 CBE's in, 184t
Methylethylether, BDE in, 168t
 bond length in, 93t
Methylethylketone, BDE in, 164t
Methylethylsulfide, BDE in, 166t
Methylethylsulfone, BDE in, 162t
Methyl formate, BDE in, 161t
2-Methylhexane, AE of, 181t
3-Methylhexane, AE of, 181t
Methyl iodide
 BDE in, 168t
 bond length in, 93t
Methyl mercaptan
 BDE in, 166t
 bond length in, 93t
2-Methylpentane

AE of, 181t
 BDE in, 165t
3-Methylpentane, AE of, 181t
Methylperoxide, BDE in, 167t
Methylphenylether, BDE in, 167t
Methylphenylketone, BDE in, 164t
2-Methyl-3-phenylpropane, BDE in, 162t
Methylphenylsulfone, BDE in, 162t
1-Methylpropanol, BDE in, 164t
2-Methylpropanol, BDE in, 164t
Methylisopropylether, BDE in, 165t
Methyl n-propyl ketone, BDE in, 164t
Methyl isopropyl ketone, BDE in, 164t
Methyl propyl sulfide, BDE in, 165t
Methyl isopropyl sulfide, BDE in, 165t
Methyl isopropyl sulfone, BDE in, 162t
Methyl radical, RE of, 168t
Methyl sulfonyl radical, RE of, 162t
Methyl thio radical, RE of, 166t
Monohalides, unstable, bond energy in, 146
Multiplicity factor, 46, 105

N

Neopentane, AE of, 181t
Neutron, 26
Nitric acid
 AE of, 139t
 BDE in, 161t
 representation of, 140f
ortho-Nitric acid, AE of, 139t
Nitric oxide
 AE of, 128t
 bond in, 132
Nitroethane, BDE in, 161t
Nitrogen
 physical state of, 73
 selected properties of, 55, 67t
Nitrogen dioxide
 AE of, 128t
 bonding in, 131
 representation of, 131f
Nitrogen fluorides, AE of, 147t
Nitrogen–hydrogen compounds, AE of,
 113t
Nitrogen oxides, bonding in, 131
Nitrogen trichloride, AE of, 150t
Nitromethane, BDE in, 161t
1-Nitropropane, BDE in, 161t
2-Nitropropane, BDE in, 161t
Nitro radical, RE of, 161t
Nitrosyl bromide, BDE in, 164t

Nitrosyl fluoride, BDE in, 164t
Nitrosyl radical, RE of, 164t
Nitrous acid, gaseous, AE of, 139t
ortho-Nitrous acid, gaseous, AE of, 139t
Nitrous oxide
  AE of, 128t
  BDE in, 164t
  bonding in, 131
  representation of, 131f
Nitryl chloride, BDE in, 161t
Nitryl fluoride, BDE in, 161t
Nonane isomers, 180
Nonmetals, physical states of, 69
Nonmolecular solids
  alternative model of bonding in, 105, 109
  bonds in, 105
Nonpolar energy, data for calculating, 100t

O

Orbital vacancies, 34
Organic chemistry
  bond energy in, 179
  uniqueness of, 178
Organic compounds, 177
Organometallic compounds, bond energy in, 189
Oxides
  diagnosis of bond type in, 124, 128t
  effect of partial charge on oxygen in, 124, 126t
  ionic bonding in, correction factor for, 135
  solid, AE of, 134, 135t
  volatile, 125
Oxyacids, 139t
Oxychloride radical, RE of, 173t
Oxyfluoride radical, RE of, 173t
Oxygen
  AE of, 19
  atom of, 17
  bond energy of, 19, 124
  bonding by, 123
  chemistry of, 123
  effect of partial charge on, 124, 126t
  partial charge on, periodicity of, 127f
  selected properties of, 55, 67t
Oxygen fluorides, AE of, 148t
Oxyhalides, AE of, 155t
Ozone, AE of, 128t, 132

P

Partial charge
  calculated from bond energy, 103t
  calculated from electronegativity, 103t
  data for calculation from electronegativity, 78t
  definition of, 9, 77
  determination of, 9, 77
  effect of, 79, 117, 124, 143
Pauli principle, 27
n-Pentane
  AE of, 181t
  BDE in, 165t
Iso-pentane, AE of, 181t
n-Pentadecane, AE of, 181t
1,4-Pentadiene, BDE in, 163t
Pentene-1, BDE in, 163t
Perchloric acid, AE of, 139t
Periodicity
  of charge on chlorine, 145f
  of charge on hydrogen, 121f
  of charge on oxygen, 127f
  of electronegativity, 42f
  of halide bond energies, 153f
  of halides, 144
  of homonuclear bond energygy, 50f
Periodic law, 2
Peroxide radical, RE of, 173t
Phenol, BDE in, 170t
Phenoxy radical, RE of, 173t
Phenyl benzyl ketone, BDE in, 161t
Phenyl bromide
  BDE in, 170t
  CBE's in, 183t
Phenylbutane, BDE in, 162t
Phenylchloride, BDE in, 170t
Phenylethylketone, BDE in, 161t
Phenyliodide, BDE in, 170t
Phenyl mercaptan, BDE in, 170t
Phenylmethylamino radical, RE of, 173t
Phenylmethylketone, BDE in, 161t
Phenylpropane, BDE in, 170t
Phenyl n-propyl ketone, BDE in, 161t
Phenyl radical, RE of, 170t
Phenyl thio radical, RE of, 173t
Phosphine, AE of, 113t
Phosphorus
  physical state of, 73
  selected properties of, 57, 67t
Phosphorus bromides, AE of, 151t
Phosphorus chlorides, AE of, 150t
Phosphorus fluorides, AE of, 147t, 148t

Phosphorus hydrogen compounds, AE of, 113t
Phosphorus iodide, AE of, 152t
Phosphorus oxides, 128t, 132
Phosphorus oxybromide, AE of, 155t
Phosphorus oxychloride, AE of, 155t
Phosphorus oxyfluoride, AE of, 155t
Phosphorus thiochloride, AE of, 155t
Polar covalence, calculation of energy of, 97
Polar covalent model, 97
Potassium, selected properties of, 59, 67t
Potassium bromide
  gaseous, AE of, 151t
  solid, AE of, 154t
Potassium chloride, 7, 15
  gaseous, AE of, 11, 104t, 150t
  solid, AE of, 15, 154t
Potassium fluoride
  gaseous, AE of, 147t
  solid, AE of, 104t, 153t
Potassium hydride, AE of, 113t
Potassium iodide
  gaseous, AE of, 104t, 152t
  solid, AE of, 155
Potassium oxide, AE of, 135t
Promotional energy, 125, 127, 151
Propane
  AE of, 181t
  BDE in, 165t
n-Propanol
  BDE in, 165t
  bond length in, 93t
Propene, BDE in, 163t
Propenylbenzene, BDE in, 162t
Propionaldehyde, BDE in, 161t
Propionate radical, RE of, 173t
Isopropoxy radical, RE of, 173t
Isopropyl acetate, BDE in, 165t
Isopropyl alcohol, BDE in, 164t
Isopropyl aldehyde, BDE in, 161t
n-Propyl allyl sulfone, BDE in, 165t
Isopropyl allyl sulfone, BDE in, 165t
n-Propyl amine, BDE in, 165t
Isopropyl amine, BDE in, 165t
n-Propylbenzene, BDE in, 162t
n-Propyl bromide
  BDE in, 165t
  bond length in, 93t
Isopropyl bromide, BDE in, 165t
n-Propyl chloride
  BDE in, 165t
  bond length in, 93t
Isopropyl chloride, BDE in, 165t

Propylene, BDE in, 169t
n-Propyl iodide, BDE in, 165t
Isopropyl iodide, BDE in, 165t
n-Propyl mercaptan, BDE in, 165t
Isopropyl mercaptan, BDE in, 165t
n-Propyl phenyl ketone, BDE in, 165t
n-Propyl radical, RE of, 165t
Isopropyl radical, RE of, 165t
Protonic bridging, 114
  energy of, 115t

Q

Quantum numbers, 27
Quantum theory, limitations of, 96

R

Radicals, reorganization of, 157
Radius
  calculation of, 10
  charged atoms, 79
  18-shell elements, 83, 90t
  hypothetical ions, 107
  "ionic", 107
  nonpolar covalent, 34
  periodicity of, 35
  relation of, to electronegativity and bond energy, 5
  reliability of, 92
  van der Waals, 35
  X-ray, 107
Radius factors, 82t
Radius ratios, 83
Reorganizational energies (RE), 157, 186
  explanations of, 171
  of molecular addition compounds, 175
Repulsion coefficient, 14, 101, 109
Resonance, 32
Rubidium, selected properties of, 62, 67t
Rubidium bromide
  gaseous, AE of, 104t, 151t
  solid, AE of, 154t
Rubidium chloride
  gaseous, AE of, 150t
  solid, AE of, 154t
Rubidium fluoride
  gaseous, AE of, 147t
  solid, AE of, 153t
Rubidium hydride, AE of, 113t
Rubidium iodide

gaseous, AE of, 152t
   solid, AE of, 104t, 155t
Rubidium oxide, AE of, 135a

S

Screening constants, 33, 81
Selenium, selected properties of, 61, 67t
Selenium (II) chloride, AE of, 150t
Selenium hexafluoride, AE of, 148t
Selenium monoxide, gaseous, AE of, 128t
Silane, AE of, 113t
Silicon, selected properties of, 57, 67t
Silicon chlorides, AE of, 150t
Silicon dioxide
   gaseous, AE of, 118t
   polymer, AE of, 135t, 137
   representations of monomer and polymer,
      138f
Silicon fluorides, AE of, 147t
Silicon hydride, AE of, 113t
Silicon monoxide
   gaseous, AE of, 128t, 130
   representation of, 130f
Silicon tetrabromide
   AE of, 151t
   crowding in, 149
Silicon tetrachloride, AE of, 104t
Silver, selected properties of, 66, 67t
Silver iodide, AE of, 103t
Sodium, selected properties of, 56, 67t
Sodium bromide
   gaseous, AE of, 151t
   solid, AE of, 154t
Sodium chloride
   gaseous, AE of, 150t
   solid, AE of, 104t, 154t
Sodium fluoride
   gaseous, AE of, 147t
   solid, AE of, 104t, 153t
Sodium hydride, AE of, 113t
Sodium iodide
   gaseous, AE of, 152t
   solid, AE of, 104t, 155t
Sodium oxide, AE of, 135t
Solids, bonding in, 13, 101, 109
Stability ratio, 41
Standard bond energies, 184, 185t
Stannane, AE of, 113t
Stibine, AE of, 113t
Strontium, selected properties of, 62
   67t

Strontium bromide, solid, AE of, 154t
Strontium chloride, solid, AE of, 154t
Strontium fluoride, AE of, 147t, 153t
Strontium oxide
   gaseous, AE of, 128t
   solid, AE of, 135t
Styrene
   BDE in, 169t
   CBE's of, 183t
Sulfur
   physical state, 58, 72
   selected properties of, 58, 67t
Sulfur chlorides, AE of, 150t
Sulfur difluoride, AE of, 149
Sulfur dioxide
   AE of, 103t
   representation of, 133f
Sulfur fluorides, AE of, 148t
Sulfuric acid
   AE of, 139t
   representation of, 141f
Sulfur (II) iodide, AE of, 152t
Sulfur monoxide, representation of, 133f
Sulfurous acid, AE of, 139t
Sulfur oxides
   AE of, 128t
   hydration of, 134t
Sulfur trioxide, representation of, 133f
Sulfuryl chloride, AE of, 155t

T

Tellurium, selected properties of, 63, 67t
Tellurium hexafluoride, AE of, 148t
Tellurium monoxide, AE of, 128t
2,2,3,3-Tetramethylbutane, BDE in, 163t
Thallium, selected properties of, 65, 67t
Thallium (I) bromide, AE of, 104t
Thallium (I) chloride, AE of, 154t
Theory, advantages of new, 24, 95, 98, 204
Thiocarbonyl radical, RE of, 161t
Thioethers, bond length in, 93t
Thiol radical, RE of, 167t
Thiols, bond lengths in, 93t
Thionyl bromide, AE of, 155t
Thionyl chloride, AE of, 155t
Thionyl fluoride, AE of, 155t
Tin, selected properties of, 63, 67t
Tin (IV) bromide, AE of, 151t
Tin (IV) chloride, AE of, 150t
Tin dioxide, AE of, 135t
Tin hydride, 113t
Tin monofluoride, AE of, 147t

Tin monoxide, AE of, 128t
Titanium, selected properties of, 66, 67t
Toluene, BDE in, 162t
Transitional elements, 65
Trihydroxytriborontrioxide, AE of, 139t
Trimethylamine, bond length in, 93t
  CBE's of, 183t
2,2,3-Trimethylbutane
  AE of, 181t
  BDE in, 163t
Triphenylmethyl radical, RE of, 173t

                            V

Vacancies, 34, 38
Valence, 31
Vinyl chloride, BDE in, 169t
Vinyl halides, 187
Vinyl radical, RE of, 169t

                            W

Water
  AE of, 103t, 113t, 128t

  BDE in, 156, 169t
  from ethane burning, 199
  protonic bridging in, 115t
  representation of, 129f

                            Z

Zinc, selected properties of, 60, 67t
Zinc bromide
  gaseous, AE of, 151t
  solid, AE of, 154t
Zinc chloride
  gaseous, AE of, 150t
  solid, AE of, 103t, 154t
Zinc fluoride
  gaseous, AE of, 147t
  solid, AE of, 103t, 153t
Zinc iodide
  gaseous, AE of, 152t
  solid, AE of, 155t
Zinc oxide, AE of, 109t, 135t
Zinc selenide, AE of, 109t
Zinc sulfide, AE of, 109t
Zinc telluride, AE of, 109t

Physical Chemistry

A Series of Monographs

Editor: **Ernest M. Loebl**

Department of Chemistry

Polytechnic Institute of New York

Brooklyn, New York

1 W. Jost: Diffusion in Solids, Liquids, Gases, 1952
2 S. Mizushima: Structure of Molecules and Internal Rotation, 1954
3 H. H. G. Jellinek: Degradation of Vinyl Polymers, 1955
4 M. E. L. McBain and E. Hutchinson: Solubilization and Related Phenomena, 1955
5 C. H. Bamford, A. Elliott, and W. E. Hanby: Synthetic Polypeptides, 1956
6 George J. Janz: Thermodynamic Properties of Organic Compounds — Estimation Methods, Principles and Practice, Revised Edition, 1967
7 G. K. T. Conn and D. G. Avery: Infrared Methods, 1960
8 C. B. Monk: Electrolytic Dissociation, 1961
9 P. Leighton: Photochemistry of Air Pollution, 1961
10 P. J. Holmes: Electrochemistry of Semiconductors, 1962
11 H. Fujita: The Mathematical Theory of Sedimentation Analysis, 1962
12 K. Shinoda, T. Nakagawa, B. Tamamushi, and T. Isemura: Colloidal Surfactants, 1963
13 J. E. Wollrab: Rotational Spectra and Molecular Structure, 1967
14 A. Nelson Wright and C. A. Winkler: Active Nitrogen, 1968
15 R. B. Andersons Experimental Methods in Catalytic Research, 1968; Volumes II and III, in preparation
16 Milton Kerker: The Scattering of Light and Other Electromagnetic Radiation, 1969
17 Oleg V. Krylov: Catalysis by Nonmetals — Rules for Catalyst Selection, 1970
18 Alfred Clark: The Theory of Adsorption and Catalysis, 1970
19 Arnold Reisman: Phase Equilibria: Basic Principles, Applications, Experimental Techniques, 1970
20 J. J. Bikerman: Physical Surfaces, 1970
21 R. T. Sanderson: Chemical Bonds and Bond Energy, 1970; Second Edition, 1976
22 S. Petrucci, ed.: Ionic Interactions: From Dilute Solutions to Fused Salts (In Two Volumes), 1971

23 A. B. F. Duncan: Rydberg Series in Atoms and Molecules, 1971

24 J. R. Anderson: Chemisorption and Reactions on Metallic Films, 1971

25 E. A. Moelwyn-Hughes: Chemical Statics and Kinetics of Solution, 1971

26 Ivan Draganic and Zorica Draganic: The Radiation Chemistry of Water, 1971

27 M. B. Huglin: Light Scattering from Polymer Solutions, 1972

28 M. J. Blandamer: Introduction to Chemical Ultrasonics, 1973

29 A. I. Kitaigorodsky: Molecular Crystals and Molecules, 1973

30 Wendell Forst: Theory of Unimolecular Reactions, 1973

31 Jerry Goodisman: Diatomic Interaction Potential Theory. Volume 1, Fundamentals, 1973; Volume 2, Applications, 1973

32 Alfred Clark: The Chemisorptive Bond: Basic Concepts, 1974

33 Saul T. Epstein: The Variation Method in Quantum Chemistry, 1974

34 I. G. Kaplan: Symmetry of Many-Electron Systems, 1975

35 John R. Van Wazer and Ilyas Absar: Electron Densities in Molecules and Molecular Orbitals, 1975

In Preparation

Brian L. Silver: Irreducible Tensor Methods: An Introduction for Chemists

A
B
C
D
E
F
G
H
I
J